A Practical Guide to OpenSees

Frontier Research in Computation and Mechanics of Materials and Biology

ISSN: 2315-4713

Published:

Vol. 4 *A Practical Guide to OpenSees*
by Quan Gu & Surong Huang

Vol. 3 *Multiscale Optimization and Materials Design*
by Jun Yan & Gengdong Cheng

Vol. 2 *Advanced Continuum Theories and Finite Element Analyses*
by James D Lee & Jiaoyan Li

Vol. 1 *Introduction to Practical Peridynamics: Computational Solid Mechanics Without Stress and Strain*
by Walter Herbert Gerstle

Frontier Research in Computation and Mechanics of Materials and Biology – Vol. 4

A Practical Guide to OpenSees

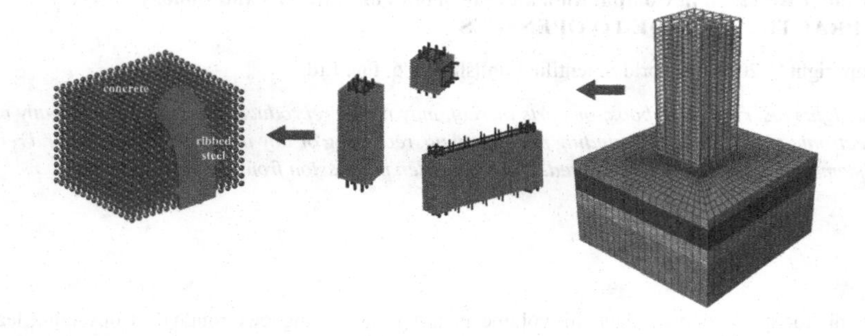

Quan Gu
Surong Huang
Xiamen University, China

World Scientific

NEW JERSEY · LONDON · SINGAPORE · BEIJING · SHANGHAI · HONG KONG · TAIPEI · CHENNAI · TOKYO

Published by

World Scientific Publishing Co. Pte. Ltd.
5 Toh Tuck Link, Singapore 596224
USA office: 27 Warren Street, Suite 401-402, Hackensack, NJ 07601
UK office: 57 Shelton Street, Covent Garden, London WC2H 9HE

Library of Congress Control Number: 2023020201

British Library Cataloguing-in-Publication Data
A catalogue record for this book is available from the British Library.

Frontier Research in Computation and Mechanics of Materials and Biology — Vol. 4
A PRACTICAL GUIDE TO OPENSEES

ISBN 978-981-120-914-7 (hardcover)
ISBN 978-981-120-915-4 (ebook for institutions)
ISBN 978-981-120-916-1 (ebook for individuals)

For any available supplementary material, please visit
https://www.worldscientific.com/worldscibooks/10.1142/11522#t=suppl

Desk Editors: Balasubramanian Shanmugam/Steven Patt

Typeset by Stallion Press
Email: enquiries@stallionpress.com

Printed in Singapore

Preface

OpenSees (the Open System for Earthquake Engineering Simulation) is an open-source finite element analysis software that can be used to model the earthquake response of structures and geotechnical systems. This framework has been under development by the Pacific Earthquake Engineering Research Center since 1997 through the U.S. National Science Foundation engineering and education centers program. As a relatively new finite element analysis software, OpenSees is dedicated to strong nonlinear analysis, with rich nonlinear elements, material libraries, and algorithms developed for strongly nonlinear geotechnical and structural systems.

The most outstanding advantage of OpenSees may be that the source code is open-developed and shared by the academic community, making it easy to achieve in-depth scientific research cooperation among scholars. Researchers do not need to repeat the program code based on the journal articles, avoiding complicated and cumbersome coding work. It has greatly improved the efficiency of scientific research. Moreover, OpenSees continuously integrates the latest scientific advances in, for example, new material models, element models, and algorithms. It brings together the latest achievements in seismic research in academia. Researchers provide a very rich and important resource library.

OpenSees has other remarkable features. It has an advanced object-oriented program architecture based on C++, which allows developers to take advantage of existing classes. It embeds sensitivity, reliability, and optimization analysis algorithms and has high-performance parallel computing capabilities with cloud through Open Science Grid, TerraGrid, and others. OpenSees has its own academic community consisting of engineers, professors, and students worldwide. It has forums, regular organized academic discussions, and training programs. OpenSees can collaborate with other systems and be used in hybrid testing through OpenFresco and other techniques. Over the years, OpenSees has made significant contributions to the academic community as an open research platform. Researchers can directly use the most cutting-edge academic results and more easily achieve academic breakthroughs. At present, there are many researchers and research teams worldwide regarding OpenSees as one of their main research platforms.

This book mainly introduces the basic applications and programming methods of OpenSees. It provides a quick introduction for readers who are expected to be researchers or engineers in civil engineering or related fields. The text has two parts. Part I introduces the various applications of OpenSees. It firstly explains how to download and run the OpenSees software, illustrates the basic syntax of the tool command language (Tcl) by using examples. It then provides various application examples, including beam–column frame structures, reinforced concrete shear walls, soil–structure interaction, fluid–solid coupling, soil liquefaction analysis, response sensitivity and reliability analyses, numerical optimization, bond-based peridynamic modeling, wheel–rail interaction for high-speed railway, and integration of OpenSees with other software based on CS technique. Part II moves on to OpenSees programming, i.e., downloading, compiling, and building OpenSees source code, as well as extending OpenSees by adding a new material model, a new element model, and a state-based peridynamics model. The examples provided are intended to be simple and easy to understand, allowing users to get started quickly. At the same time, the examples strive for realism although on a very small scale, e.g., the material parameters, boundary conditions, and loading conditions are nevertheless similar to those of realistic large models. Users are expected to be able to model realistic complex engineering problems by extending the examples in this book.

This book is based on the authors' lectures at Xiamen University over the past several years. Most of the examples in this book can be found on the website: https://github.com/OpenSeesXMU. Much of this book's content derives from feedback and long-term support of OpenSees users, which is highly appreciated. We also would like to express our heartfelt thanks to Dr. Lei Wang for the massive composition, review, and correction work, particularly the checking of the codes on the github website line by line. We also want to thank Ning Zhang, Jinghao Pan, Baoyin Sun, and Yongdou Liu for their review work. We are especially grateful to Professor Jinping Ou for his long-term support for the authors' work on OpenSees. A technical book of this length may always have deficiencies. It will be much appreciated if readers can provide feedback to help us further improve the quality of the book.

About the Authors

Quan Gu is a Professor of Civil Engineering at Xiamen University of China. He received his Bachelor's and Master's degrees from Tsinghua University of China and got his PhD degree from the University of California, San Diego (UCSD), in 2008. He joined Xiamen University in 2010. Dr. Gu has kept involved in the OpenSees development and promotion in China. He has participated in a few major research projects which are related to the further development of OpenSees or conducted based on OpenSees. He has published more than 70 academic papers and served as guest editor for several journals including *Soil Dynamics and Earthquake Engineering*, *International Journal of Structural Stability and Dynamics*, *Computer Modeling in Engineering & Sciences*, and *Buildings*.

Surong Huang is currently Assistant Professor at Xiamen University of China. She graduated with a PhD from the University of California, San Diego (UCSD), and Bachelor's and Master's degrees from Tsinghua University of China.

Contents

Part I

Introduction to OpenSees Application

Chapter 1

Download and Run

Step 1: Go to OpenSees website http://opensees.berkeley.edu/ and click DOWNLOAD on the left menu bar to enter the download page (Fig. 1.1). Users need to register, and it is free. Click registration as shown in Fig. 1.2. That takes you to the homepage shown in Fig. 1.3 where you need to click "I agree to these terms" to continue to the next step.

Step 2: On the page shown in Fig. 1.4, fill in the basic information and click "submit". Then login to the mailbox to activate it.

Step 3: Return to the OpenSees homepage shown in Fig. 1.2, enter the registered mail, and click "submit".

Step 4: Download the OpenSees file (assuming you are using Windows), as shown in Fig. 1.5. Unzip the .zip file and find OpenSees.exe in one of the subfolders. Double click OpenSees.exe and run it as shown in Fig. 1.6.

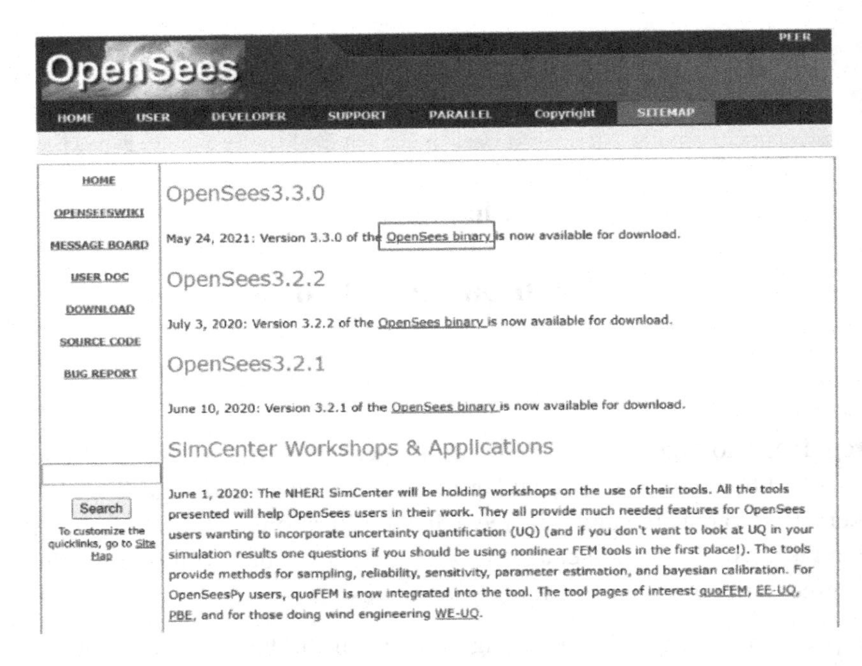

Fig. 1.1 The OpenSees website.

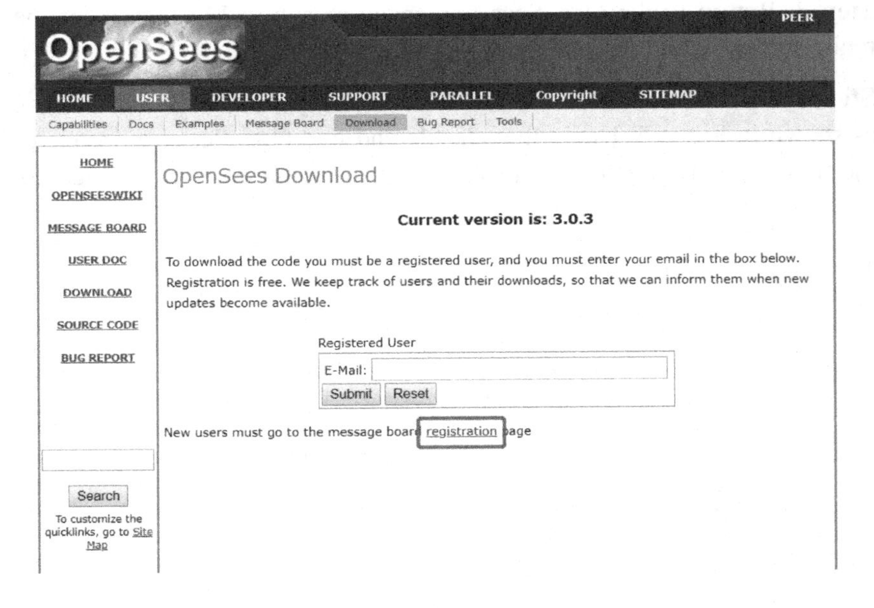

Fig. 1.2 E-mail registration (1).

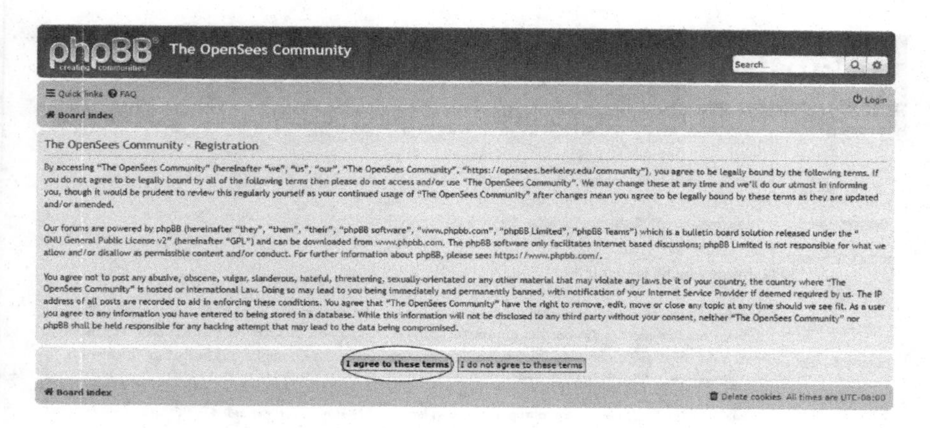

Fig. 1.3 E-mail registration (2).

The OpenSees Community - Registration

Please note that you will need to enter a valid e-mail address before your account is activated. You will receive an e-mail

Username:
Username must be between 3 and 20 chars long and use only alphanumeric characters.

E-mail address:

Confirm e-mail address:

Password:
Password must be between 6 and 100 characters long, must contain letters in mixed case and must contain numbers.

Confirm password:

Language: British English

Timezone: [UTC - 8] Pacific Standard Time

The items marked with * are required profile fields and need to be filled out.

What is the middle number?: *
1 2 3

CONFIRMATION OF REGISTRATION

Identify element types that might appear in a structural model from the list: tower bee beam failure shear steel concrete shell:
This question is a means of preventing automated form submissions by spambots.

Reset Submit

Fig. 1.4 E-mail registration (3).

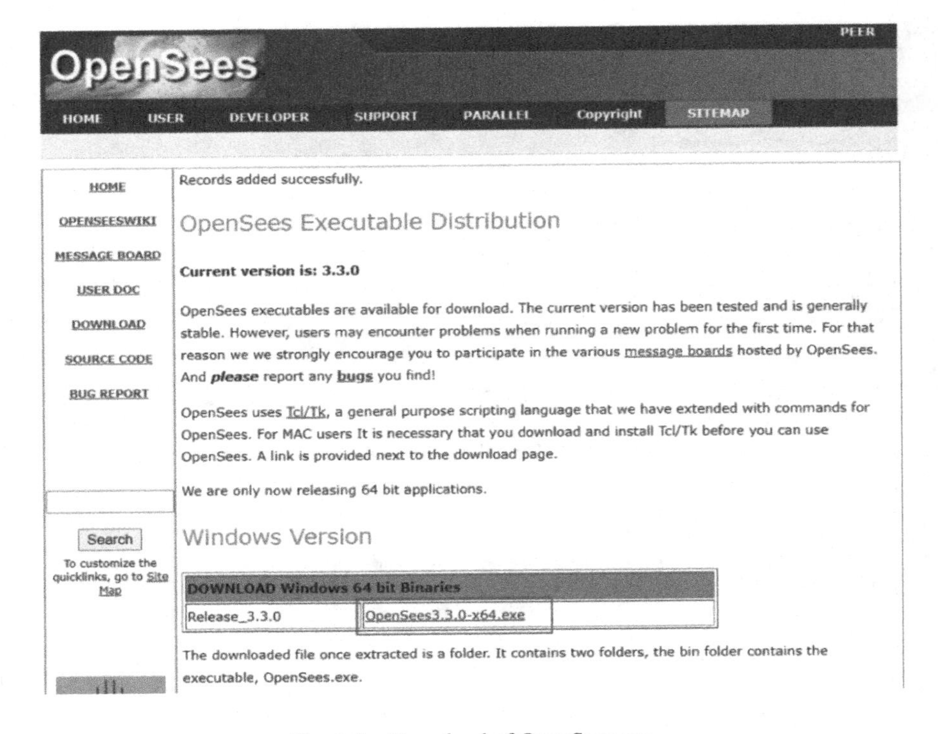

Fig. 1.5 Download of OpenSees.exe.

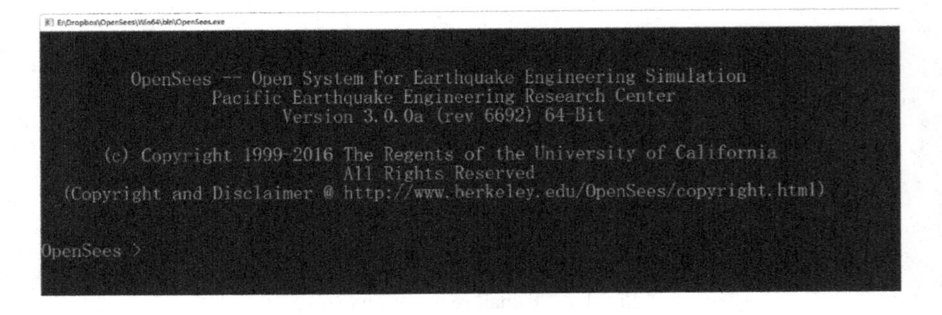

Fig. 1.6 Run OpenSees.exe.

Chapter 2

A Simple Structural Analysis Example

This chapter presents a simple example of creating a truss system model and performing static and dynamic analyses using OpenSees (Fig. 2.1). A special tool command language (Tcl) is employed to build the model. A detailed description about Tcl will be presented in Chapter 3. For a quick study, readers are encouraged to input the following Tcl code using a text editor such as Notepad or other scripting language editor, and to save the code into a file with .tcl extension. Note that you need to remove the line numbers in the following Tcl code (and codes in other chapters as well). A simple way to run the .tcl file is to put the file in the same folder as OpenSees.exe. Otherwise you will need to set the global file path to find the .tcl file.

A typical finite element analysis (FEA) for a structural system consists of two parts: modeling and analyzing. The modeling part involves building the finite element (FE) model, which is shown in lines 1–16:

```
1   wipe;
2   model Basic -ndm 2 -ndf 2
3   if { [file exists output] == 0 } {
4               file mkdir output;
5   }
6   node 1 0.0 0.0
7   node 2 144.0 0.0
8   node 3 168.0 0.0
9   node 4 72.0 96.0
```

```
10  fix 1 1 1
11  fix 2 1 1
12  fix 3 1 1
13  uniaxialMaterial Elastic 1 3000.0
14  element truss 1 1 4 10.0 1
15  element truss 2 2 4 5.0 1
16  element truss 3 3 4 5.0 1
```

In the first line, the "wipe" command deletes all existing FE models in OpenSees, including nodes, elements, materials, boundary conditions, and so on.

> *Note*: The "*wipe*" command will not delete the variables defined in Tcl (for example, if a variable *a* has been defined by "*set a 100*", the variable *a* will still exist in the memory of Tcl after performing a "wipe"). We can take advantage of this feature in numerical optimization and other analyses, when the design variables are retained in the Tcl memory while the model can be removed (e.g., using the *wipe* command) and rebuilt.

The second line "*ndm*" denotes the number of dimensions. The "2" indicates that the model is a two-dimensional (2D) model. "*ndf*" is the abbreviation for "number of degrees of freedom (dof)", and the second "2" indicates that each node has 2 dofs.

Lines 3–5 create a folder named "output" to save the output files (for convenience only; these lines can be removed if not necessary).

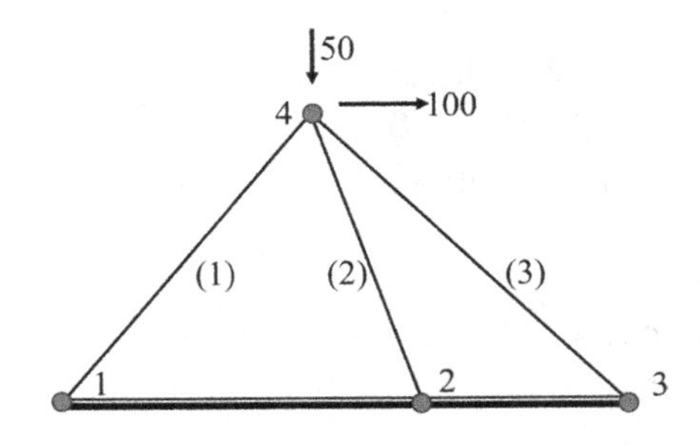

Fig. 2.1 A simple truss example.

Lines 6–9 create 4 nodes. The numbers following "node" are the node's tag and its *x* and *y* coordinates. For example, in line 6 a node at (0.0, 0.0) is defined as node #1.

Lines 10–12 define a single point (SP) constraint. The displacements in both the *x* and *y* directions of nodes 1, 2, and 3 are fixed.

Line 13 defines a material. The material type is a uniaxial elastic material with material tag 1 (for later reference), and 3000.0 is its elastic modulus. The units in OpenSees are user-specified, but they must be consistent throughout the model. This book uses the international system (IS) of units unless otherwise mentioned.

Lines 14–16 define elements. The element type is "truss", and the numbers following "truss" are the element tag, the tags of the two end nodes, the cross-sectional area, and the tag of the material previously defined. For example, in line 14 a new truss element with tag 1 is defined with two end nodes #1 and #4. The cross-sectional area of the element is 10.0, and the tag of its material is #1 (i.e., the elastic material defined in line 13).

Lines 17–20 specify the output. Unlike other finite element software, OpenSees does not save or export all information about nodes, elements, materials, etc., but only outputs the information that the user specifies.

```
17  recorder Node -file output/disp_4.out -time -node 4 -dof 1 2 disp
18  recorder Node -file output/reaction_1.out -time -node 1 -dof 1 2
    reaction
19  recorder Node -file output/reaction_2.out -time -node 2 -dof 1 2
    reaction
20  recorder Node -file output/reaction_3.out -time -node 3 -dof 1 2
    reaction
```

The "Node" after the keyword "recorder" means that the node's information is recorded. There are other types of recording methods available such as Element and Graphics. "-file" denotes saving the information to a file. For example, line 17 records the response of node 4 in a file named "disp_4.out" in the folder "output". "-time" means that the time is recorded as the first column. By default the responses at every time step are recorded. "-node 4" means the information of node #4 is recorded. "-dof 1 2" indicates recording its first and second dofs (i.e., in the x and y directions). "disp" indicates that the information to be recorded is the node's displacement. Similarly, lines 18–20 record the reactions of nodes #1, #2, and #3 (indicated by using the keyword "reaction") in the x and y directions.

The external forces are of three types: node forces, base excitation, and multiple support. The forces are defined using the pattern command. In this example, an external force pattern is applied to node #4:

```
21  pattern Plain 1 Linear {
22  load 4 100.0 -50.0
23  }
```

Line 21 defines a pattern tagged 1. "Linear" means that the external force is increasing linearly. The actual force applied to node 4 at each step is the current system time (to be dealt with later) multiplied by the external force coefficient defined in line 22 (i.e., the coefficient on the 4th node: $F_x = 100$, $F_y = -50$). In other words, the real force on the node #4 is $F_x = 100 * t$, $F_y = -50 * t$ (where t is the current system time), rather than those defined in line 22. The negative sign indicates that the force direction is opposite to the specified positive direction.

Line 23 completes the modeling for this simple example. We now turn to the analysis. The Tcl commands for static analysis are as follows:

```
24  constraints Transformation
25  numberer RCM
26  system BandSPD
27  test NormDispIncr 1.0e-6 6 2
28  algorithm Newton
29  integrator LoadControl 0.1
30  analysis Static
31  analyze 10
```

Line 24 shows how the boundary constraint equations are handled in a finite element model. In addition to "Transformation", "Plain" and "Penalty" are available. For simple single point (SP) constraints, (e.g., using "fix" commands) all these methods work well. For multi-point (MP) constraints (e.g., using the "equalDOF" command), only "Plain" and "Penalty" work.

Line 25 shows the numbering method for the structure's dof. In addition to RCM, there are AMD law, Plain law, and others available.

Line 26 shows how the equation is stored and solved. "BandSPD" works only for cases in which the stiffness matrix is symmetric, but alternative methods are available.

Line 27 indicates the methods used to judge whether the convergence (e.g., of the Newton algorithm for solving the equation of motion) is achieved at each time step. Here "NormDispIncr" is used: when the norm of the incremental displacement is less than 1.0e-6, the iteration is considered to converge. The number "6" denotes that the maximum iteration number allowed at each time step is 6, otherwise the algorithm cannot converge in that step. The last number denotes whether the calculation process should be displayed on the DOS screen (2 indicates that convergence information will be displayed only after convergence of each time step, rather than at each iteration of a step).

Line 28 specified that Newton's algorithm is used for solving the equation of motion. There are other algorithms available: a modified Newton algorithm, a Newton line search algorithm, and a BFGS algorithm. It is possible to switch algorithms during the analysis in case of non-convergence.

Lines 29–31 represent the loading method. "LoadControl" means the load control method is applied. (There are other control methods including displacement control, arc-length control, and central difference, depending on whether static or dynamic analysis is being performed.) 0.1 means that at each step the load factor increment is 0.1 times the external force coefficient (specified in the "pattern" command in Line 21, note the system time increases by 0.1 s in each step); "static" means that static analysis is performed; "analyze 10" specifies analyzing 10 times or loading steps (i.e., the system time is 1.0 s after the static analysis).

After completing the calculation, you can check the displacement of node #4. Just type "nodeDisp" at the OpenSees prompt as follows:

```
OpenSees > nodeDisp 4 1
        0.53009277713228364000
OpenSees > nodeDisp 4 2
        -0.17789363846931766000
```

That gives the node displacement in the x and y directions, respectively. You can also check the output directory to find the files storing displacement and bearing reactions.

> *Note*: In this example, the load factor increment 0.1 is combined with the total of 10 time steps to specify the loading method. In each of the 10 time or loading steps, the incremental forces are $\Delta F_x = 0.1 * 100 = 10$; $\Delta F_y = 0.1 * (-50) = -5$. In other words, the external force increases linearly. In more general cases, the external force is not necessarily linearly increasing, and a "Series" command can be used. For example:
>
> *Pattern Plain 1 { Series –time { 0.0 1.0 2.0 3.0}–values*
> *{0.0 1.0 0.0 -1.0} } { load 4 100.0 -50.0}*
> *integrator LoadControl 0.1*
> *analysis Static*
> *analyze 30*
>
> That defines the force series shown in Fig. 2.2. Although it is a static analysis, OpenSees keeps the "time" as in a dynamic analysis. The time step size is 0.1 s, so after 30 steps the system time is 3.0 s.

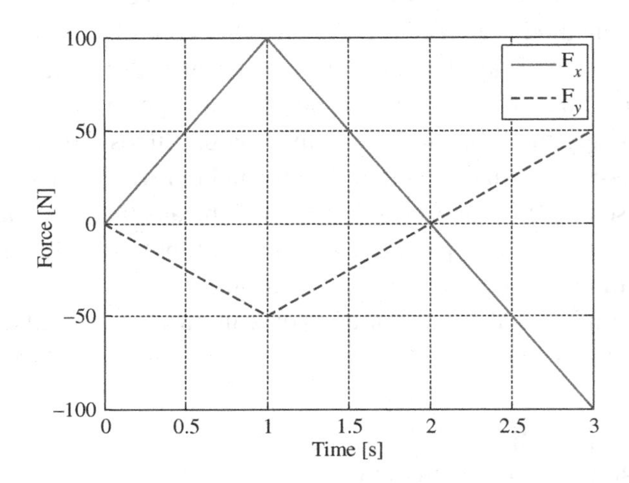

Fig. 2.2 The force series.

Note: In OpenSees, the Tcl command "nodeReaction" works only if a "recorder" command has been defined to record the node's reaction (refer to Line 18). If not, it gives incorrect results.

In OpenSees, the dynamic analysis command flow is very similar to that for static analysis. To perform dynamic analysis only using this example (i.e., without static analysis), you can use lines 32–43 to replace lines 24–31, or just add a "reset" command after line 31 to reset the model to its initial state before typing lines 32–43. In the current example, dynamic analysis is performed after the static analysis, i.e., performing following commands after line 31 (this situation is common, such as applying gravity followed by seismic analysis):

```
32  wipeAnalysis
33  loadConst -time 0.0
34  mass 4 100.0 100.0
35  pattern UniformExcitation 2 1 -accel ''Series -factor 3 -filePath
    elcentro.txt -dt 0.01''
36  constraints Transformation
37  numberer RCM
38  system BandSPD
39  test NormDispIncr 1.0e-6 6 4
40  algorithm Newton
41  integrator Newmark 0.5 0.25
42  analysis Transient
43  analyze 2000 0.01
```

Line 32 removes all analysis commands (i.e., lines 24–31) while preserving the FE model of lines 1–23. (Note that a simple "wipe" command will remove all modeling and analysis in lines 1–31, while "wipeAnalysis" will not remove modeling).

Line 33 means that all the previously defined external forces in the static analysis so far will be kept unchanged. In this example, it means that $F_x = 100 * 0.1 * 10 = 100$ and $F_y = -50 * 0.1 * 10 = -50$ will keep unchanged. "*-time 0.0*" resets the current time from 1.0 (i.e., $0.1 * 10$) seconds to 0.

> *Note*: The "loadConst" in line 33 is very important. It ensures that the external force (defined as pattern #1 in line 21) will keep unchanged from now on, otherwise it will continue increasing linearly. "-time 0.0" resets the time to 0.0 s to continue the subsequent dynamic analysis (or system time will be 1.0 s).

Line 34 defines the mass of node 4, which is 100.0 in both the x and y directions. (OpenSees allows them to be different).

Line 35 applies a seismic load as the base excitation. The number "2" is the tag assigned to that load pattern (just as load pattern #1 defined in Line 21). The new load pattern does not necessarily have to be numbered #2, just any number different from #1. The number "1" means the acceleration is in the x direction (2 for the y direction, etc.). The content in the double quotes (i.e., "Series..") defines an acceleration series whose values are read from the file "elcentro.txt" multiplied by a factor of 3.0, while the time interval between two consecutive accelerations in the file is 0.01 s. The data in "elcentro.txt" should be in float format with one column only.

Lines 36–40 are the same as those in the static analysis.

Line 41 specifies that Newmark's implicit method is used with two parameters γ and β equal to 0.5 and 0.25, respectively.

Line 42 denotes that transient analysis will be performed.

Line 43 performs the analysis (2000 time steps) with a time step size of 0.01 s.

After the analysis is complete, the calculation results (the first ten static analysis results) are stored in the file disp_4.out, where the first and second columns are time and horizontal displacements (according to line 17). The displacements of the transient analysis will be found after the 10th row of the disp_4.out file, since the data in the first 10 rows are static responses. The predicted horizontal displacement of node 4 is plotted in Fig. 2.3.

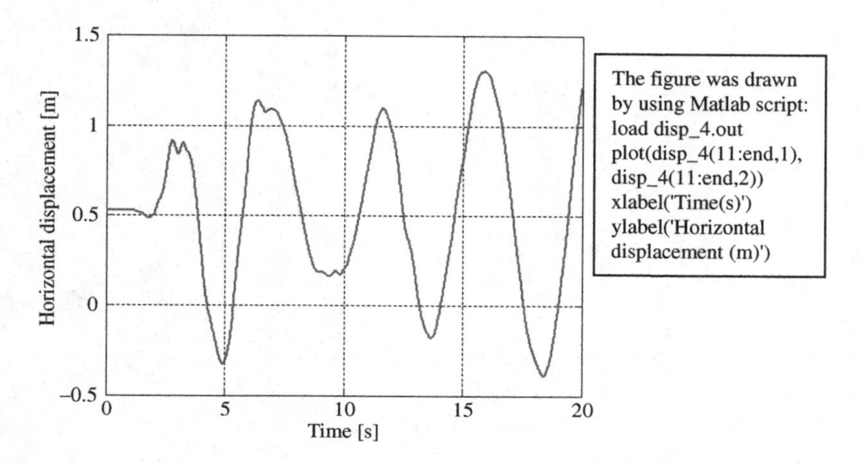

Fig. 2.3 Predicted horizontal displacement of node 4.

Note: The 0.01 s specified in line 43 is not necessarily the same as that defined in the time series (line 35). If the two time intervals are not the same, OpenSees will automatically get the acceleration at each integration time step (defined in line 43) by linear interpolation of the acceleration series (defined in line 35).

Chapter 3

Introduction to Tcl Syntax

In OpenSees, the pre-processing is based on a language called Tool Command Language (Tcl) (https://www.tcl.tk/). Tcl is not suitable for post-processing, and there are optional software available, e.g., Matlab, GID (https://www.gidsimulation.com/, http://gidopensees.rclab.civil.auth.gr/), OpenSees Navigator (https://openseesnavigator.berkeley.edu/), etc. This chapter introduces the basic Tcl syntax. Users may input the following Tcl examples (i.e., after the prompt OpenSees>) to quickly learn Tcl. You may skip this chapter if you do not need to write complex Tcl code.

3.1 Tool Command Language and OpenSees

Tcl is an advanced general-purpose programming language, created by John Ousterhout and co-workers. Tcl has many advantages, e.g., simplicity and power, which make it suitable as a pre-processing tool for OpenSees. The relationship between the operating system (OS), Tcl, and OpenSees is shown in Fig. 3.1.

The Tcl serves as a command interpreter on top of the OS. In this sense, OpenSees extends Tcl language by adding new commands for building models and FE analysis, e.g., "node", "element", "analyze".

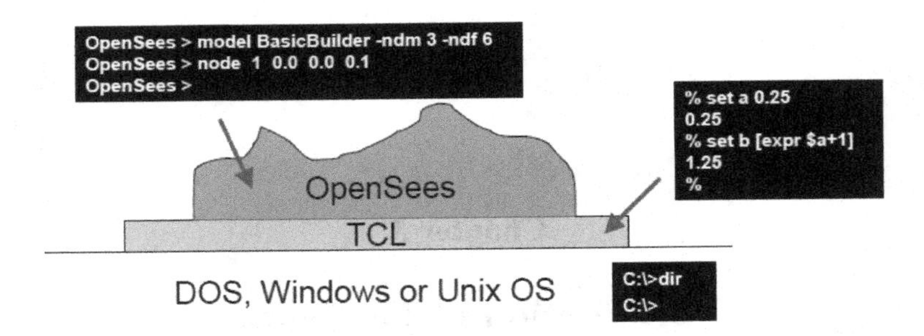

Fig. 3.1 The relationship between the operating system, Tcl, and OpenSees.

3.2 Basic Syntax

The basic format of a Tcl command is

command arg1 arg2 ... ;

The first word "command" is the command's name, while arg1, arg2 and so on are the command's parameters. Spaces are used to separate the different parameters. The ";" denotes the end of this command, and can be omitted if there is nothing after it. Here are explanations for some basic and commonly used Tcl commands.

Command	Explanation
set	Set a value to a variable. For example, assign the value 2.0 to the variable "A": OpenSees> set A 2.0
unset	Delete one or more variables from Tcl memory.
expr	Pass the argument to an expression interpreter and return the evaluated result
puts	Output text
info exists	Check if the variable exists
info global	Return a list of the names of all the global variables
$	Variable substitution symbol
[]	Command substitution symbol.
#	Comment symbol. "#" is usually used at the beginning of a comment line

3.2.1 *Example 1: Basic Tcl syntax and special characters*

- *Input*

```
set E 1
set I 1
set EI [expr $E *$I]
puts ‘‘$EI\n’’
puts {$EI\n}
```

- *Output*

```
>
1
$EI\n
```

Note: “{}” cancels the substitution function of $.

3.2.2 *Example 2: Unset and info exists*

- *Input*

```
set m 2
unset m                ; # Delete the variable ‘‘m’’
if {![info exists m]}  ; # Check if the variable ‘‘m’’ exists.
 {set m 0
}else {
  set m [expr $m+1]
}
puts ‘‘$m’’
```

- *Output*

```
>
0
```

3.2.3 *Example 3: Info global*

- *Input*

```
info global            ; # List all global variables in the memory
```

- *Output*

```
>
tcl_rcFileName tcl_version argv argv0 tcl_interactive E auto_path
errorCode EI errorInfo auto_execs auto_index I env tcl_patchLevel m
argc n tcl_library tcl_platform
```

- *Input*

```
puts "$tcl_version"        ; #Query the current Tcl version
```

- *Output*

```
>
8.5
```

3.3 Variables

A simple Tcl variable contains the variable name and its value, which are case sensitive. There is no need to declare a variable before using it. A Tcl array is a container of variables.

- Array operation commands

Command	Explanation
array exists arrayName	Determine whether an array exists, returns 1 if the array exists, returns 0 otherwise
array get arrayName	Return a list of array values
array size arrayName	Return the size of the array
array set arrayName datalist	Define an array and initialize it
array unset arrayName	Delete an array, release memory

3.3.1 *Example 4: Set an array*

- *Input*

```
set data(1) 0.5                    ; # data(1) = 0.5
set data(2) 1.4                    ; # data(2) = 1.4
puts [array get data]              ; # get the data array
set a $data(1)                     ; # a = data(1)
set data(2) 1.5                    ; # data(2) = 1.5
puts [expr $a+ $data(2)]           ; # c = a+data(2)
```

- *Output*

```
>
1 0.5 2 1.4
2.0
```

- *Input*

```
set data_1(name) William           ; # array can be more flexible
set data_1(age) 23
set data_1(gender) male
set data_1(occupation) work
set size_data_1 [array size data_1]
puts "$size_data_1"
puts "[array get data_1]"
array unset data_1                 ; # Delete the array, and release
                                         memory
if {[array exists data_1] == 0} {  ; # check if the array exists
 puts "data_1 is not an array"
}else {
 puts "data_1 is an array"
}
```

- *Output*

```
>
4
occupation worker gender male age 23 name William
data_1 is not an array
```

3.4 Expressions

The Tcl script is a set of sequentially executed commands that consist of expressions and statements. A Tcl expression usually combines operands, operators, and parentheses.

- The following operators are used:

Operator	Explanation
n*m, n/m, n%m	n*m: n multiplied by m; n/m: n divided by m; n%m: remainder
n+m, n-m	n+m: n plus m; n-m: n minus m
!n	!n: logical NOT. if n is 0, then return 1, otherwise return 0;
n>m, n<m	n>m: If n is greater than m, return 1, otherwise, return 0; n<m: If n is less than m, return 1, otherwise, return 0
n>=m, n<=m	n>=m: If n is greater than or equal to m, return 1, otherwise, return 0; n <= m: If n is less than or equal to m, return 1, otherwise, return 0
n==m, n!=m	n==m: If n is equal to m, return 1, otherwise, return 0; n! = m: If n is not equal to m, return 1, otherwise return 0

○ Mathematical functions available are as follows:

Function	Explanation
abs(n)	The absolute value function
acos(n)	The arc cosine function
cos(n)	The cosine function
asin(n)	The arc sine function
sin(n)	The sine function
log(n)	The natural logarithm function
log10(n)	The base 10 logarithm function
sqrt(n)	The square root function
exp(n)	The exponential function

3.4.1 *Example 5: Operators and functions*

• *Input*

```
set M 64
set N 81
set W [expr "$M + $N"]
set W [expr $M + $N]
set W_LABEL "$M plus $N is"
puts "$W_LABEL $W"
puts "The square root of $M is [expr sqrt($M)]\n"
```

• *Output*

```
>
64 plus 81 is 145
The square root of 64 is 8.0
```

3.5 String Operations

Tcl string operations include calculating the lengths of strings, comparing strings, replacing a string, etc.

- Basic string manipulation commands are as follows:

Command	Explanation
string length	Calculate the number of characters in a string
string compare	Compare the strings character by character. Returns −1, 0, or 1, depending on whether the first string is lexicographically less than, equal to, or greater than the last one.
string equal	Compare the strings character by character. Returns 1 if the two strings are identical, or 0 when not.
append	Strings are appended to the value stored in a variable
format	Produce a formatted string from a given template.

3.5.1 *Example 6: String operations*

- *Input*

```
set number_1 2016/
set number_2 04/16
set number_3 [append number_1 $number_2]
set Length_1 [string length $number_1]
set Length_2 [string length $number_2]
set Length_3 [string length $number_3]
puts ‘‘$number_1’’
puts ‘‘$number_2’’
puts ‘‘$number_3’’
puts ‘‘$Length_1’’
puts ‘‘$Length_2’’
puts ‘‘$Length_3’’
```

```
if {[string compare $Length_1 $Length_2] == 0} {
     puts {Length_1 and Length_2 are equal}
} else {
puts {Length_1 and Length_2 are not equal}}
if {[string equal $number_1 $number_3]} {
     puts {number_1 and number_3 are equal}
} else {
     puts {number_1 and number_3 are not equal}
}
```

- *Output*

```
>
2016/04/16
04/16
2016/04/16
10
5
10
Length_1 and Length_2 are not equal
number_1 and number_3 are equal
```

3.6 Lists

The list is one of the basic data structures in Tcl. It is used to represent an ordered collection of elements.

o Basic list command explanations are as follows:

Command	**Explanation**
list arg1 arg2 ...	Create a new list of the arguments
lindex list i	Return the value of the (i+1)th parameter of the list
llength list	Return the number of elements in the list

3.6.1 *Example 7: List operations*

- *Input*

```
set name_list {liming wangfang chenxiaoli};
# Create a list of names
set age_list {23 42 13};
# Create a list of ages
set gender_list {male female female};
# Create a list of genders
set data_list [list name_list age_list gender_list];
# Create a new list containing three lists
set name_list_length [llength $name_list];
# [] returns the number of elements in ''name_list''
set data_list_length [llength $data_list];
# [] returns the number of elements in ''data_list''
puts ''$name_list_length''
puts ''$data_list_length''

lindex $datalist 1;
# Return the second element of ''data_list''
```

- *Output*

```
>
3
3
age_list
```

3.7 Control Structure

There are Tcl commands which implement control structures. The looping commands — while, for, foreach — are one type. The conditional commands such as if and switch are another.

o The basic control command formats are as follows:

Command	Basic format
if	if (condition){statement_1} else if{ statement_2} else{statement_3}
if_else	
if_elseif_else	
switch	switch flags value { pattern1 body1 pattern2 body2 ··· }
while	While (condition) { Statement}
for	for {initial} {test} {final} {statement}
foreach	foreach Var list {statement}

3.7.1 *Example 8: For and foreach*

• *Input*

```
set sum_1 0
for {set n 0} {$n<6} {incr n} {
    set sum_1 [expr $sum_1 + $n]
}
set sum_2 0
set list [list 1 2 3 4 5]
foreach m $list {
    set sum_2 [expr $sum_2+$m]
}
puts ‘‘sum_1 is $sum_1’’
puts ‘‘sum_2 is $sum_2’’
if {$sum_1 < $sum_2} {                ; # Note there is a space in the
                                              middle of it ‘‘} {’’
    puts ‘‘$sum_2 is greater than $sum_1’’
}elseif {$sum_1 > $sum_2} {           ; # Notice the blank before and
                                              after ‘‘elseif’’
    puts ‘‘$sum_1 is greater than $sum_2’’
} else {
    puts ‘‘$sum_1 and $sum_2 are equal’’
}
```

• *Output*

```
>
sum_1 is 15
sum_2 is 15
15 and 15 are equal
```

3.7.2 *Example 9: While and switch*

• *Input*

```
set n 0;
set sum 0;
while {$n < 5} {incr n;set sum [expr $sum + $n]}
puts ‘‘sum is $sum’’
switch -- $sum_1 {
     15 {puts ‘‘sum is $sum’’}
     default {puts ‘‘sum is not $sum’’}
}
```

• *Output*

```
>
sum is 15
sum is 15
```

3.8 Procedure

The proc command is used to create a new procedure. The basic format is as follows:

proc procName argList body

where “procName” is the name of the procedure, “argList” specifies the procedure’s arguments, and “body” is the process block containing the Tcl script.

3.8.1 *Example 10: Proc*

- *Input*

```
proc sum {num_1 num_2 num_3} {     ; # Create a new procedure
      set m [expr $num_1 + $num_2 + $num_3];
      return $m
}
set a 1
set b 2
set c 3
puts ''The sum of $a + $b + $c is [sum $a $b $c]''
```

- *Output*

```
>
The sum of 1 + 2 + 3 is 6
```

3.9 File Commands

File is a Tcl command that can manipulate file names and attributes.

o Explanations of file command

Command	Explanation
cd	Change the current directory to the given one
pwd	Return the current directory
file mkdir	Create new directories
file delete	Remove files and directories
file copy	Make a copy of the file
file exists	Return 1 if the file exists, or 0 otherwise

3.9.1 *Example 11: File mkdir, file copy, and file delete*

- *Input*

```
file mkdir Data                    ; # Create a directory
logFile Data/data.txt          ; # Create a file in the folder
if {[file exists Data] == 0} {
        file mkdir Data;
        logFile Data/data.txt
} else {
        file copy Data Data_1
        file delete Data_1/data.txt       ;
}
cd Data_1                        ; # Change current directory to "Data_1"
pwd                              ; # Show the current directory
```

- *Output*

```
>
E:/OpenSees book/tcl/Data_1
```

This chapter has introduced the very basic Tcl syntax, and users can build an FE model and analyze it by combining these existing tcl commands with the OpenSees Tcl commands that will be introduced later. The following chapters will provide some examples of how to use OpenSees to model and analyze structures step by step. Users are recommended to type these Tcl scripts and run these examples for quick study purpose.

Chapter 4

Frame Structure Analyses

4.1 Example 1: Static and Dynamic Analyses for a Two-Dimensional (2D) Elastic Reinforced Concrete Column

(A) Example description

This chapter uses an elastic reinforced concrete (RC) column as an example. The details are shown in Fig. 4.1. The elastic moduli and the cross-sectional areas of each element are identical. The elastic modulus is 3.0×10^{10} Pa and the size of the cross-section is 0.5 m × 0.5 m. The models in this book use SI units unless otherwise mentioned. Two types of analyses will be performed.

(1) Gravity and pseudo-static pushover analyses. The maximum horizontal pushover displacement is 0.5 m.
(2) Earthquake base excitation with a maximum acceleration of 0.90 g ($g = 9.8\ m/s^2$).

(B) Tcl commands

Gravity and pseudo-static pushover analyses

```
1 wipe
2 model basic -ndm 2 -ndf 3
3 if {[file exists output] == 0} {
4    file mkdir output;
5 }
```

```
6 node 1 0 0
7 node 2 0 3.0
8 node 3 0 6.0
9 fix 1 1 1 1
10 geomTransf Linear 1
11 element elasticBeamColumn 1 1 2 0.25 3.0e10 5.2e-3 1
12 element elasticBeamColumn 2 2 3 0.25 3.0e10 5.2e-3 1
```

As the Tcl commands have already been introduced in the previous chapters, they will not be explained again in this chapter.

Line 2 indicates it is a 2D model and each node in the model has 3 degrees of freedoms (dofs), i.e., horizontal, vertical, and rotational dofs.

Line 9 represents constraints. The horizontal, vertical, and rotation dofs of node #1 are constrained.

Line 10 performs coordinate transformation. It defines a "linear" coordinate transformation from the basic to the global coordinate system. (The "linear" indicates that large displacement is not considered. Other coordinate transformations will be introduced later). The transformation tag is 1. The details of the three coordinate systems — basic, local, and global — will be presented in Section 4.2. The global and local coordinate systems in this model are shown in Fig. 4.2.

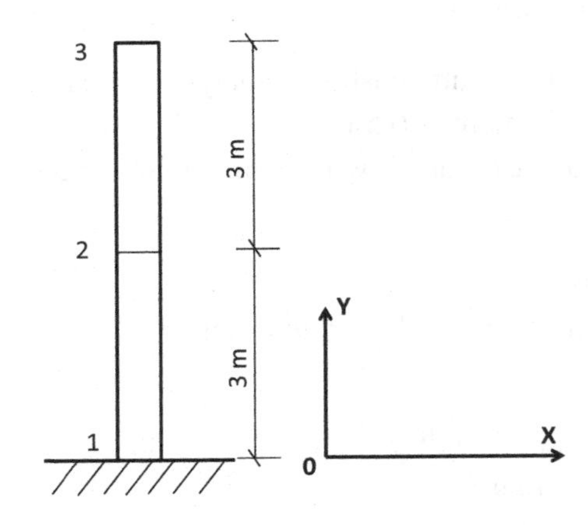

Fig. 4.1 An example of elastic cantilever.

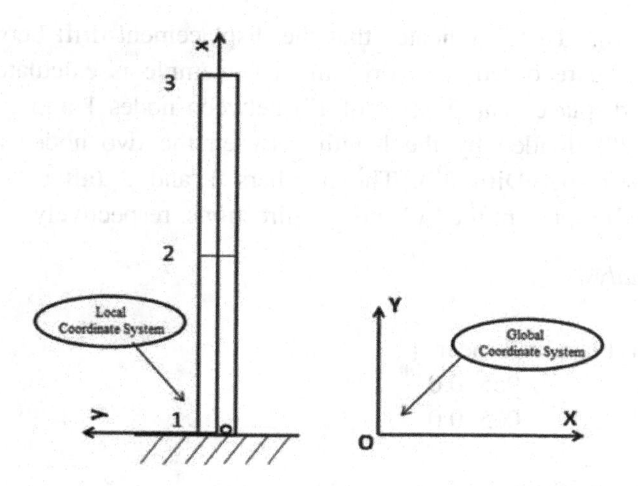

Fig. 4.2 The global and local coordinate systems of the elastic column model.

Lines 11 and 12 create 2 elastic beam elements. The numbers after "elasticBeamColumn" represent the tag of an element, tags of two element end nodes, the cross-sectional area, Young's modulus, the moment of inertia, and the number of the coordinate transformation previously defined, respectively.

```
13 recorder Node -file output/disp_3.out -time
   -node 3 -dof 1 2 3 disp
14 recorder Node -file output/disp_2.out -time
   -node 2 -dof 1 2 3 disp
15 recorder Node -file output/reaction_1.out
   -time -node 1-dof 1 2 3 reaction
16 recorder Drift -file output/drift_1.out -time
   -iNode 1-jNode 2 -dof 1 -perpDirn 2
17 recorder Drift -file output/drift_2.out -time
   -iNode 2-jNode 3 -dof 1 -perpDirn 2
18 recorder Element -file output/force_1.out
   -time -ele 1 globalForce
```

These lines are the recorders.

In line 16, "Drift" indicates that the displacement drift between two nodes will be recorded. The drift in this example is calculated as the horizontal displacement (i.e., "-dof 1") between nodes 1 and 2 ("-iNode 1 -jNode 2") divided by the length between the two nodes along the *y* direction ("-perpDirn 2"). The numbers 1 and 2 (after "-dof" and "-perpDirn") represent the "*x*" and "*y*" directions, respectively.

Gravity analysis

```
19 pattern Plain 1 Linear {
20    load 2 0. -1.0e5 0.0
21    load 3 0. -1.0e5 0.0
22 }
23 constraints Plain
24 numberer Plain
25 system BandGeneral
26 test NormDispIncr 1.0e-8 6 2
27 algorithm Newton
28 integrator LoadControl 0.1
29 analysis Static
30 analyze 10
31 puts ''Gravity analysis is finished...''
```

The above lines of code perform the gravity analysis.

> *Note*: For static elastic analysis, the results are identical when the gravity load is applied in 1 loading step or in 10 sequential smaller steps. However, the results may be different for static gravity analysis of large elastoplastic structures such as high-rise buildings or geotechnical systems. They usually require step-by-step loading to avoid non-convergence caused by excessive loading in a single step.

Lines 20 and 21 indicate that a force coefficient (not force!) of -1.0×10^5 is applied to both nodes 2 and 3 in the *y* direction. In this example, the increment of loading force at each step is the force coefficient multiplied by 0.1 s defined in Line 28 (i.e., -1.0×10^4). Therefore, it takes 10 steps to complete the gravity analysis (Line 30).

Horizontal pushover analysis

The following code performs the pushover analysis.

```
32 loadConst -time 0.0
33 pattern Plain 2 Linear {
34     load 2 0.5 0.0 0.0
35     load 3 1.0 0.0 0.0
36 }
37 integrator DisplacementControl 3 1 0.001
38 analyze 500
39 puts ''Horizontal force pushover analysis
   is completed.''
```

Line 32 keeps the gravity loads unchanged and resets the time to 0.0 s.

Lines 33–35 define a pattern with tag 2. The force is applied to both nodes 2 and 3 in the x direction, with a loading ratio of 0.5:1 between them.

Line 37 specifies the loading method, i.e., a displacement control will be used to apply loads. A horizontal ("1" indicates the global x direction) displacement increment of 0.001 m is applied to node 3 at each time or loading step.

Line 38 performs a 500-step pushover analysis. The final horizontal displacement of node 3 is 0.5 m.

Analysis results: The displacement response to the maximum distributed force (the force on node 3) is shown in Fig. 4.3 (plotted using Matlab). The output file "disp_3.out" records the "time" and the horizontal displacement of node 3 in columns 1 and 2, respectively. Note that the "time" is not actually the time in the displacement control case, but it is the same as the maximum force herein.

> *Note*: When the displacement control method is used to apply loads, the forces will be adjusted at each step according to the ratio (i.e., 0.5:1, or 1:2) defined in the pattern in lines 34 and 35 until the target displacement is reached (the horizontal displacement of node 3 prescribed in lines 37 and 38). The records in column 1 of the output file "disp_3.out" are then no longer actual "time", but a factor. The actual force is the force coefficient in pattern #2 (defined in Lines 34 and 35) multiplied by that factor.

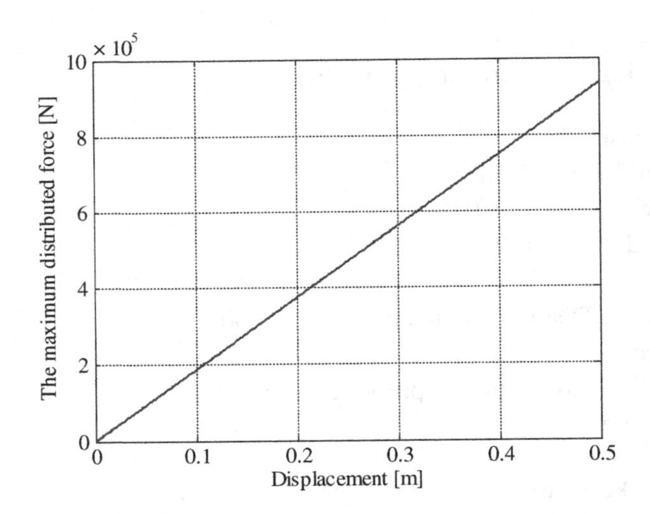

Fig. 4.3 Maximum distributed force vs. displacement.

Earthquake base excitation

Seismic analysis is usually performed after the gravity analysis, i.e., to perform the following code after line 39.

```
40 mass 2 1.0e4 0.0 0.0
41 mass 3 1.0e4 0.0 0.0
42 loadConst -time 0.0
43 timeSeries Path 1 -dt 0.02 -filePath tabas.txt
   -factor 9.8
44 pattern UniformExcitation 2 1 -accel 1
45 set temp [eigen 1]
46 scan $temp "\%e" w1s
47 set w1 [expr sqrt($w1s)]
48 puts "First order frequency f: [expr $w1/6.28]"
49 set ksi 0.02
50 set a0 0
51 set a1 [expr $ksi*2.0/$w1]
52 rayleigh $a0 0.0 $a1 0.0
53 wipeAnalysis
54 constraints Plain
```

```
55 numberer Plain
56 system BandGeneral
57 test NormDispIncr 1.0e-8 10 2
58 algorithm Newton
59 integrator Newmark 0.5 0.25
60 analysis Transient
61 analyze 1000 0.02
62 puts ''Ground motion analysis completed...''
```

Lines 40 and 41 specify the lumped mass of nodes 2 and 3, respectively. In OpenSees, node mass and element mass are calculated separately and added together later.

Line 43 defines a time series tagged as #1 for the earthquake loading data. The data are read from a file "tabas.txt" and multiplied by a load factor 9.8. The time interval between two sequential data points is 0.02 s. The data in "tabas.txt" should be in float format with one column only.

Line 44 defines a load pattern with tag 2 describing a base uniform excitation along the *x* direction. The acceleration input is taken as the time series with tag 1 previously defined. The seismic wave defined in tabas.txt is shown in Fig. 4.4.

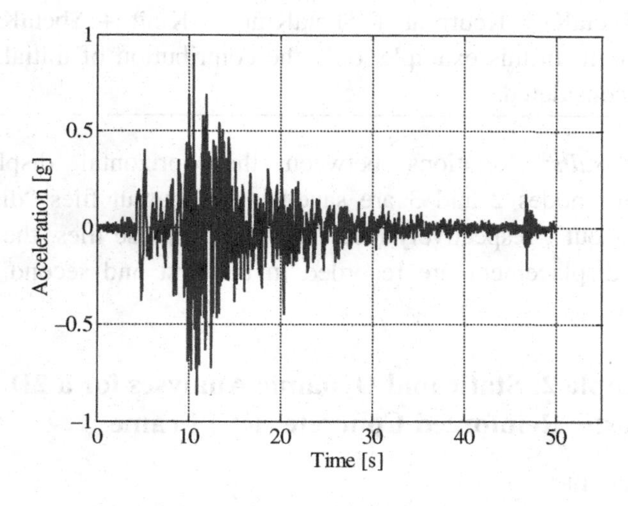

Fig. 4.4 The earthquake acceleration history.

Lines 45–52 calculate the Rayleigh damping coefficients a_0 and a_1 based on the damping ratio. In line 45, the Tcl command "eigen 1" calculates a characteristic value of the model, which is the square of the model's first natural angular frequency. The value is stored in a string-type variable "temp".

Lines 46–48 convert the "temp" string into a real-type variable "w1s", and calculate its square root "w1". Then display the model's first natural frequency on the screen.

Lines 49–51 calculate the Rayleigh damping coefficients a_0 and a_1.

Line 52 defines the Rayleigh damping in each finite element and at each node. In OpenSees, the state variables (e.g., mass, stiffness, and damping) of the nodes and elements are calculated independently and assembled at the structure level later.

> *Note*: There are four parameters in the command defining a Rayleigh damping matrix: "$alphaM", "$betaK", "$betaKinit", and "$betaKcomm". "$alphaM" is the coefficient of the mass matrix, and the other three parameters are the coefficients of the stiffness matrix in the current iteration (not necessarily a converged time step), the initial (elastic) stiffness matrix and the stiffness matrix at the last converged time step, respectively. The Rayleigh damping matrix will superimpose contributions of the mass and stiffness matrices of the form D = $alphaM * M + $betaK * Kcurrent + $betaKinit * Kinit + $betaKcomm * KlastCommit. In this example, only the contribution of initial stiffness matrix is considered.

Analysis results: Relations between the horizontal displacements and time for nodes 2 and 3 are saved in the output files "disp_2.out" and "disp_3.out", respectively (see Fig. 4.5). In those files, the time and horizontal displacement are recorded in the first and second columns, respectively.

4.2 Example 2: Static and Dynamic Analyses for a 2D Inelastic Reinforced Concrete (RC) Frame

(A) The example

The model is shown in Fig. 4.6. The cross-section of the beam is linear elastic, while those of the two columns are nonlinear. The cross-sections

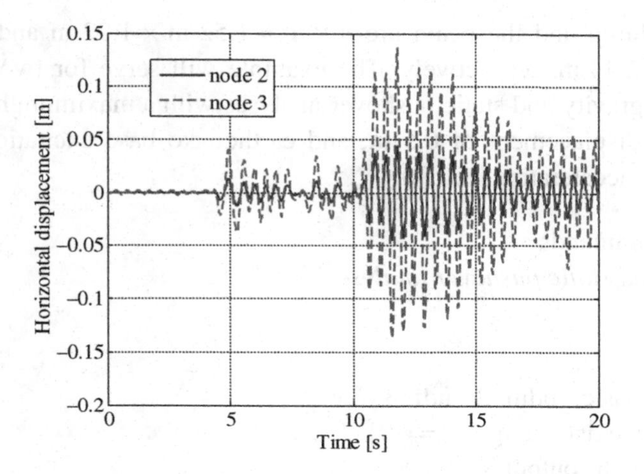

Fig. 4.5 Horizontal nodal displacement histories.

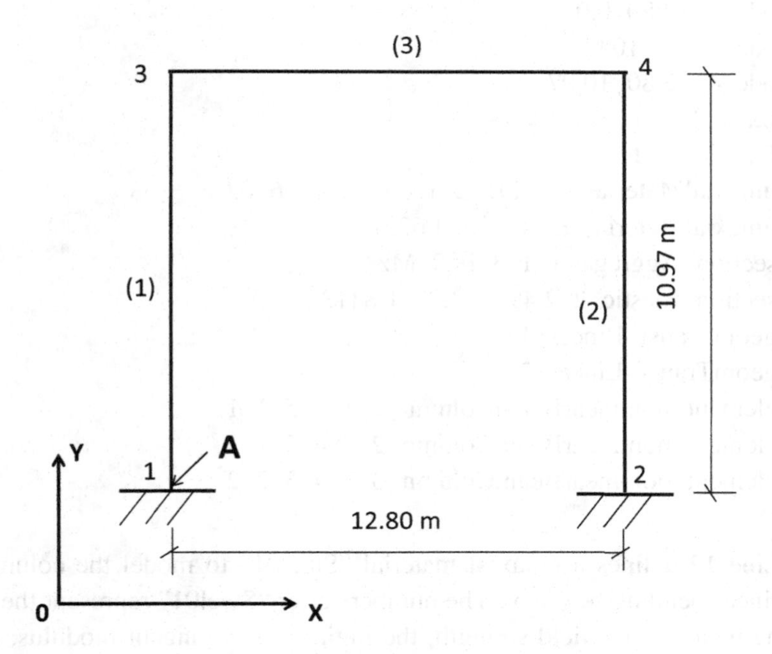

Fig. 4.6 The 2D inelastic reinforced concrete frame.

of the column and the beam are $b \times h = 1.52$ m $\times$ 1.22 m and $b \times h =$ 1.52 m $\times$ 2.44 m, respectively. The example will serve for two types of analyses: gravity and static pushover analyses with a maximum horizontal pushover displacement of 0.5 m; and earthquake base excitation with a maximum acceleration of 0.90g.

(B) Tcl commands
Gravity and static pushover analyses

```
1 wipe
2 model basic -ndm 2 -ndf 3
3 if {[file exists output] == 0} {
4   file mkdir output
5 }
6 node 1 0.0 0.0
7 node 2 12.80 0.0
8 node 3 0.0 10.97
9 node 4 12.80 10.97
10 fix 1 1 1 1
11 fix 2 1 1 1
12 uniaxialMaterial Steel01 2 1.47e4 5.74e6 0.01
13 uniaxialMaterial Elastic 3 4.62e7
14 section Aggregator 1 3 P 2 Mz
15 section Elastic 2 2.49e7 3.72 1.8413
16 geomTransf Linear 1
17 geomTransf Linear 2
18 element nonlinearBeamColumn 1 1 3 5 1 1
19 element nonlinearBeamColumn 2 2 4 5 1 1
20 element nonlinearBeamColumn 3 3 4 5 2 2
```

Line 12 defines a uniaxial material “Steel01” to model the column’s nonlinear bending behavior. The numbers after “Steel01” represent the tag of the material, its yield strength, the initial elastic tangent modulus, and the steel’s strain-hardening ratio, respectively. The constitutive behavior for the material “Steel01” is shown in Fig. 4.7.

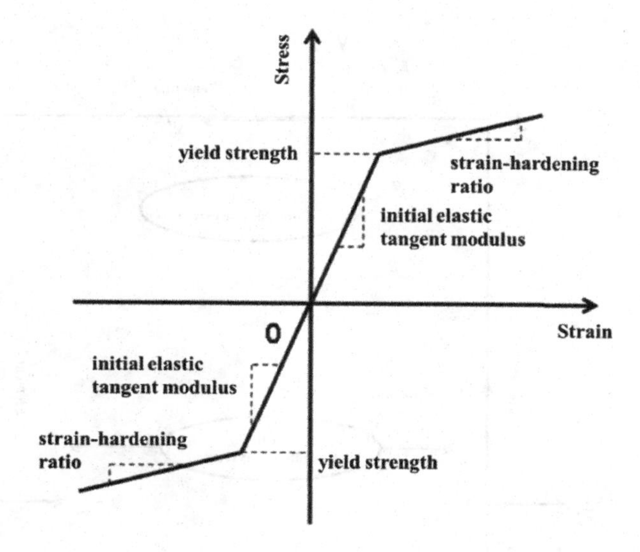

Fig. 4.7 The constitutive behavior of "Steel01".

Line 13 defines the axial characteristics of the column as uniaxial elastic. The numbers following "Elastic" represent the material's tag and its tangent (elastic modulus).

Line 14 defines an aggregated section (with tag 1) which combines the two previously defined 1D materials into a single-section force-deformation model. The letters "P" and "M_Z" indicate the axial and bending characteristics of the section, respectively. The numbers before those letters are the numbers of the previously defined materials. In this example, material #3 (defined in line 13) is used to define the axial force–deformation relationship. Correspondingly, the physical meaning of the last number in line 13 is changed from the tangent (elastic modulus) E to the tensile/compressive stiffness EA. Similarly, material #2 (in line 12) is used for the moment–curvature relationship about the local *z*-axis of the section, and the meaning of the last three numbers changes to the yield moment M_y, the bending stiffness EI, and the ratio of post-yield bending stiffness to the pre-yield bending stiffness. The local *z*-axis is outwardly perpendicular to the paper (see Fig. 4.8).

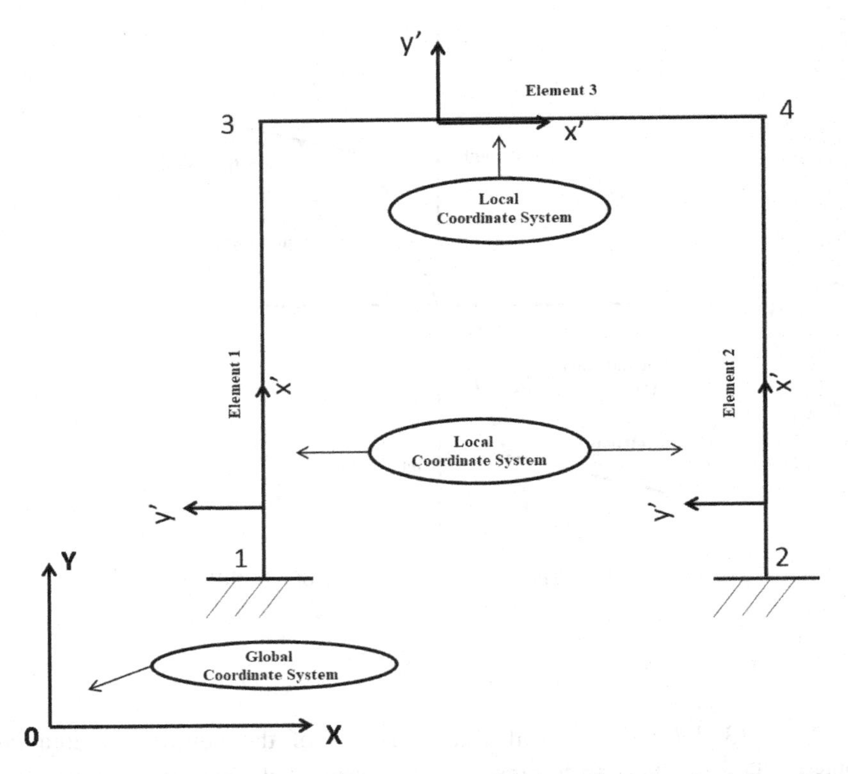

Fig. 4.8 The global and local coordinate systems of the two-dimensional reinforced concrete frame.

> *Note*: On the official OpenSees website (opensees.berkeley.edu), the material parameters of "Steel01" are Fy, E0, and b, which represent the yield strength, the initial elastic modulus, and the strain-hardening ratio assumed, respectively. However, in this example the Tcl command "section Aggregator" is used (in line 14), which changes the meanings of the parameters in lines 12 and 13 correspondingly. In line 12, the yield strength Fy, the elastic modulus E, and the strain-hardening ratio b become the yield moment My, the bending stiffness EI, and the ratio of post-yield bending stiffness to the pre-yield bending stiffness. In line 13, the elastic modulus E becomes the tensile/compressive stiffness EA.

Line 15 defines the beam's elastic section. The numbers following "Elastic" are the tag of the section, the elastic modulus E, the cross-sectional area A, and the moment of inertia of the section I_z, respectively.

Lines 16 and 17 define two "linear" coordinate transformations which transform beam element stiffness and resisting force from the basic system to the global coordinate system in a "linear" way, i.e., such that large geometric deformation is not considered. The global and local coordinate systems in this model are defined in Fig. 4.8. In 2D frame structure analysis, there is no need to specify the local coordinate direction, so we can just define one coordinate transformation in this example, e.g., line 17 can be deleted and all three elements can use the same linear coordinate transformation #1, i.e., change the last number of line 20 from 2 to 1.

> *Note*: The x direction of the local coordinate system is along the beam or column (e.g., in line 18, the local x axis of element 1 is along the vertical direction from node 1 to node 3), while the z direction is always outwardly perpendicular to the paper.

Lines 18–20 define 3 force-based nonlinear beam column elements. The numbers following "nonlinearBeamColumn" represent the element's number, the numbers of its two end nodes, the numbers of the Gauss points to be used in the Gauss–Lobatto integration, the number of the section, and the number of the previously defined coordinate transformation.

```
21 recorder Node -file output/disp_34.out -time -node 3 4 -dof 1 2 3
   disp
22 recorder Node -file output/reaction_12.out -time -node 1 2 -dof 1
   2 3 reaction
23 recorder Drift -file output/drift_1.out -time -iNode 1 2 -jNode 3 4
   -dof 1 -perpDirn 2
24 recorder Element -file output/force_12.out -time -ele 1 2 global-
   Force
25 recorder Element -file output/foce_3.out -time -ele 3 globalForce
26 recorder Element -file output/forcecolsec_1.out -time -ele 1 2
   section 1 force
27 recorder Element -file output/defocolsec_1.out -time -ele 1 2
   section 1 deformation
```

```
28 recorder Element -file output/forcecolsec_5.out -time -ele 1 2
   section 5 force
29 recorder Element -file output/defocolsec_5.out -time -ele 1 2
   section 5 deformation
30 recorder Element -file output/forcebeamsec_1.out -time -ele 3
   section 1 force
31 recorder Element -file output/defobeamsec_1.out -time -ele 3
   section 1 deformation
32 recorder Element -file output/forcebeamsec_5.out -time -ele 3
   section 5 force
33 recorder Element -file output/defobeamsec_5.out -time -ele 3
   section 5 deformation
```

Those lines of code are the necessary recorders.

The "Element" in lines 24–33 means information about an element is recorded. In lines 24 and 25, "-ele 1" indicates that the #1 element's information is recorded. "globalForce" means the global element force is recorded. In lines 26–33 "Section 1" indicates that the response of the 1st Gauss point (closest to the first end node) is recorded. "force" means that the section stress resultant in the local coordinate system is recorded, "deformation" means the section deformation in the local coordinate system is recorded.

```
34 pattern Plain 1 Linear {
35 eleLoad -ele 3 -type -beamUniform -122.5
36 }
37 constraints Plain
38 numberer Plain
39 system BandGeneral
40 test NormDispIncr 1.0e-8 6 2
41 algorithm Newton
42 integrator LoadControl 0.1
43 analysis Static
44 analyze 10
45 puts ''Gravity analysis is finished...''
```

Those lines of code perform the gravity analysis.

Line 34 defines a pattern #1 with linearly increasing load. Line 35 indicates that a uniform load of -122.5 is applied to element 3 in the local coordinate system's y direction.

```
46 loadConst -time 0.0
47 pattern Plain 2 Linear {
48     load 3 1.0 0.0 0.0
49     load 4 1.0 0.0 0.0
50 }
51 integrator DisplacementControl 3 1 0.001
52 analyze 500
53 puts ''Horizontal force pushover analysis
   is completed...!''
```

Analysis results: The displacement response to the pushover force (the force on node 3, but it is the same as that on node 4) is shown in Fig. 4.9.

The output file "disp_34.out" records the "time" in its column 1 (which is not the time in the displacement control situation, but the same as the pushover force on node 3 or node 4) and the horizontal displacement of node 3 in column 2.

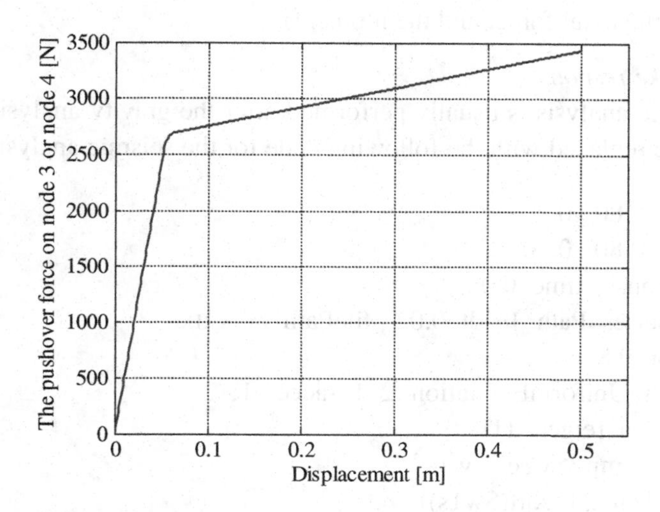

Fig. 4.9 The pushover force vs. displacement.

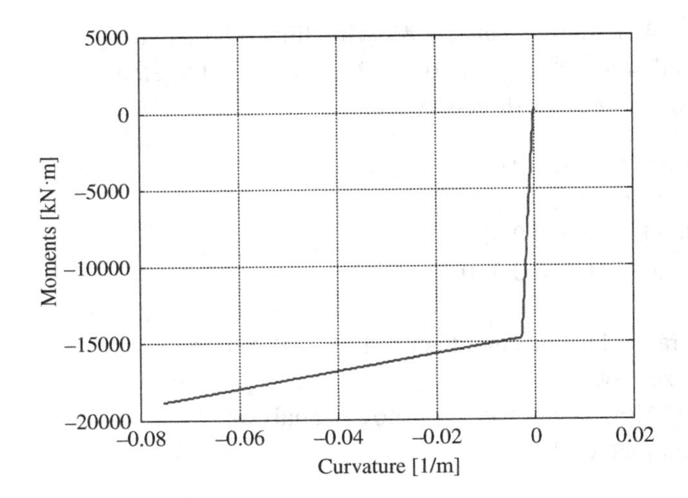

Fig. 4.10 Moment–curvature responses of Section A.

The nonlinear moment–curvature relationship for Section A of the column (Fig. 4.6) under the local coordinate system is shown in Fig. 4.10. The curvature is recorded in column 3 of the "defocolsec_1.out" file (where the 3 columns record the time, axial strain, and curvature). The moment is recorded in column 3 of "forcecolsec_1.out" (where the 3 columns present the time, the axial force, and the moment).

Earthquake response

Seismic analysis is usually performed after the gravity analysis. Lines 46–53 are replaced with the following code for the seismic analysis.

```
54 mass 3 80. 0. 0.
55 mass 4 80. 0. 0.
56 loadConst -time 0.0;
57 timeSeries Path 1 -dt 0.02 -filePath tabas.txt
   -factor 9.8;
58 pattern UniformExcitation 2 1 -accel 1;
59 set temp [eigen 1]
60 scan $temp "\%e" w1s
61 set w1 [expr sqrt($w1s)]
62 puts "First order frequency f: [expr $w1/6.28]"
```

```
63 set ksi 0.02
64 set a0 0
65 set a1 [expr $ksi*2.0/$w1]
66 rayleigh $a0 0.0 $a1 0.0
67 wipeAnalysis
68 constraints Plain
69 numberer Plain
70 system BandGeneral
71 test NormDispIncr 1.0e-8 10 2
72 algorithm Newton
73 integrator Newmark 0.5 0.25
74 analysis Transient
75 analyze 1000 0.02
76 puts ''Ground motion analysis completed...''
```

Analysis results: The horizontal displacement histories of the top node 3 are plotted in Fig. 4.11 using the output file "disp_34.out" in which the time and the horizontal displacement are recorded in the first and second columns.

The nonlinear moment–curvature relationship for Section A of the column (Fig. 4.6) in the local coordinate system is shown in Fig. 4.12.

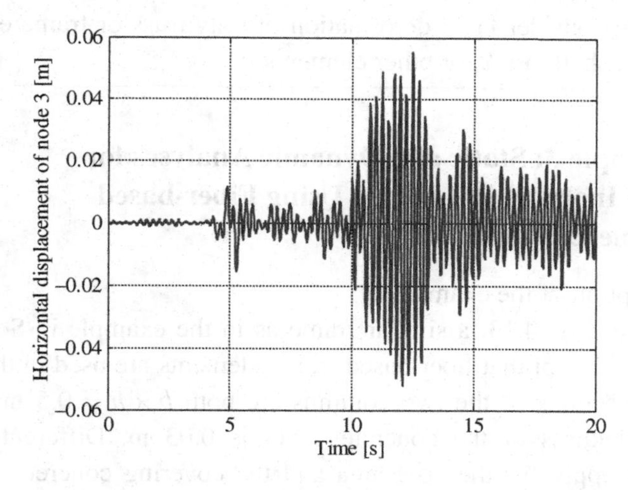

Fig. 4.11 The horizontal displacement histories of node 3.

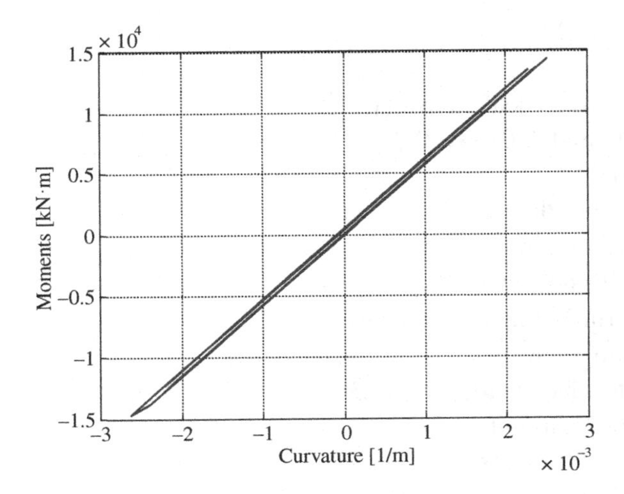

Fig. 4.12 Moment–curvature responses of Section A.

The curvature is recorded in column 3 of the “defocolsec_1.out” file, while the moment is recorded in column 3 of “forcecolsec_1.out”.

> *Note*: In this example, geometric (large) deformation can be considered if “geomTransf” is modified., i.e., the "Linear" in lines 16 and 17 is changed to “PDelta” or “Corotational”. The current version of OpenSees (3.0.3) can consider large deformation of only truss or frame elements, rather than shell, brick, or other elements.

4.3 Example 3: Static and Dynamic Analyses for a 2D Inelastic RC Frame Using Fiber-based Frame Element

(A) Description of the example

As shown in Fig. 4.13, a similar frame as in the example in Section 4.2 is analyzed, except that fiber-based frame elements are used in the model. The cross-sections of the two columns are both $b \times h = 0.5$ m $\times$ 0.5 m, and the thickness of the concrete cover is 0.03 m. Different material parameters apply for the core area and the covering concrete. Six steel bars are used here. The beam is linear elastic, and the section size is

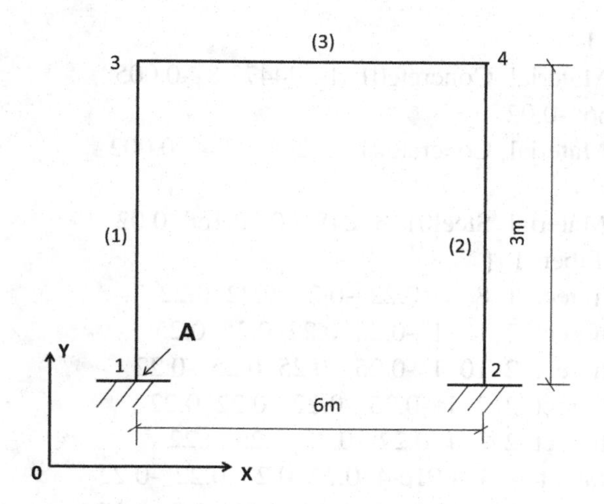

Fig. 4.13 A 2D inelastic FRC frame model.

$b \times h = 0.25$ m $\times$ 0.6 m. As before, the units are meter (m), ton (t), second (sec), and kiloNewton (kN).

Two types of analyses are to be performed:

(1) Gravity and static pushover analyses. The maximum horizontal pushover displacement is 0.3 m;
(2) Earthquake base excitation with a maximum acceleration of 0.90 g ($g = 9.8$ m/s^2).

(B) Tcl commands

Gravity and static pushover analyses

```
1 wipe
2 model basic -ndm 2 -ndf 3
3 if {[file exists output] == 0 }{
4     file mkdir output
5 }
6 node 1 0.0 0.0
7 node 2 6.0 0.0
8 node 3 0.0 3.0
9 node 4 6.0 3.0
10 fix 1 1 1 1
```

```
11 fix 2 1 1 1
12 uniaxialMaterial Concrete01 1 -34473.8 -0.005
   -24131.66 -0.02
13 uniaxialMaterial Concrete01 2 27579.04 -0.002
   0.0 -0.006
14 uniaxialMaterial Steel01 3 248200. 2.1e8 0.02
15 section Fiber 1 {
16     patch rect 1 8 8 -0.22 -0.22 0.22 0.22
17     patch rect 2 10 1 -0.25 0.22 0.25 0.25
18     patch rect 2 10 1 -0.25 -0.25 0.25 -0.22
19     patch rect 2 2 1 -0.25 -0.22 -0.22 0.22
20     patch rect 2 2 1 0.22 -0.22 0.25 0.22
21 layer straight 3 3 4.91e-4 0.22 0.22 0.22 -0.22
22 layer straight 3 3 4.91e-4 -0.22 0.22 -0.22 -0.22
23 }
24 section Elastic 2 3.0e7 0.15 4.5e-3
25 geomTransf Linear 1
26 geomTransf Linear 2
27 element dispBeamColumn 1 1 3 5 1 1
28 element dispBeamColumn 2 2 4 5 1 1
29 element dispBeamColumn 3 3 4 5 2 2
```

Lines 12 and 13 define the nonlinear uniaxial concrete material “Concrete01” (i.e., Kent–Scott–Park model) for the core concrete and the cover concrete, respectively. The numbers following “Concrete01” are the tag of the material, the compressive strength at 28 days, the strain at maximum strength, the crushing strength, and the strain at the crushing strength. Those values should be positive. If they are given negative, OpenSees will convert them to positive automatically. The constitutive behavior of the material “Concrete01” is shown in Fig. 4.14. As the figure shows, the material cannot sustain tensile stress.

Line 14 defines a uniaxial inelastic steel material “Steel01”. As has already been explained, the numbers following “Steel01” represent the tag of material, the yield strength, the initial elastic tangent modulus, and the material’s strain–hardening ratio. The constitutive behavior of “Steel01” is shown in Fig. 4.7.

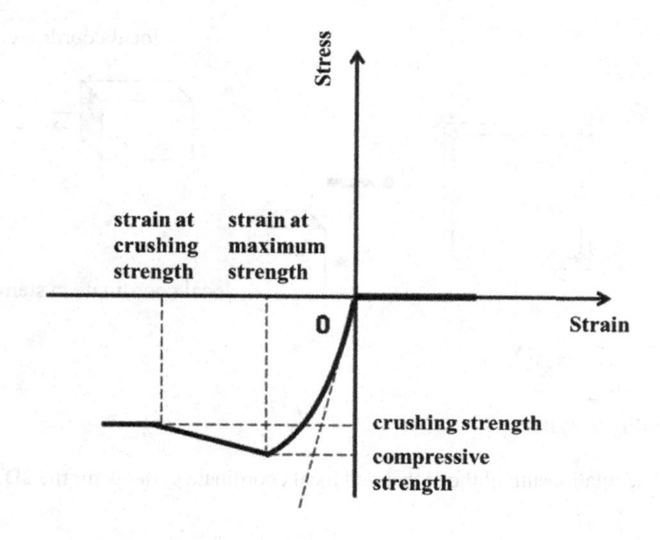

Fig. 4.14 The constitutive behavior of the material "Concrete01".

Line 15 defines a fiber cross-section with number 1. The "patch rect" in each line of 16–20 defines a rectangular fiber patch. The fiber section is a combination of all five fiber patches. The first number after "patch rect" is the tag of the material in that patch; the second and third numbers are the number of subdivisions (i.e., fibers) in the local y and z directions (to be defined later); the fourth and fifth numbers are the minimum y' and z' coordinates of the rectangle in the local coordinate system (to be defined); the sixth and seventh numbers are the rectangles' maximum y' and z' coordinates.

The "layer straight" in lines 21–22 defines the distinct fibers (in this case the steel bars) along a straight line. The numbers following "layer straight" are the tag of material, the number of bars along the line, the bars' cross-sectional area, the y' and z' coordinates of the first steel bar, the y' and z' coordinates of the last steel bar, respectively.

Lines 27–29 define displacement beam column elements. The numbers following "dispBeamColumn" are the tag of the element, the tags of its two end nodes, the number of Gauss points to be used in the Gauss–Lobatto integration, the tags of the section and the previously defined coordinate transformation, respectively.

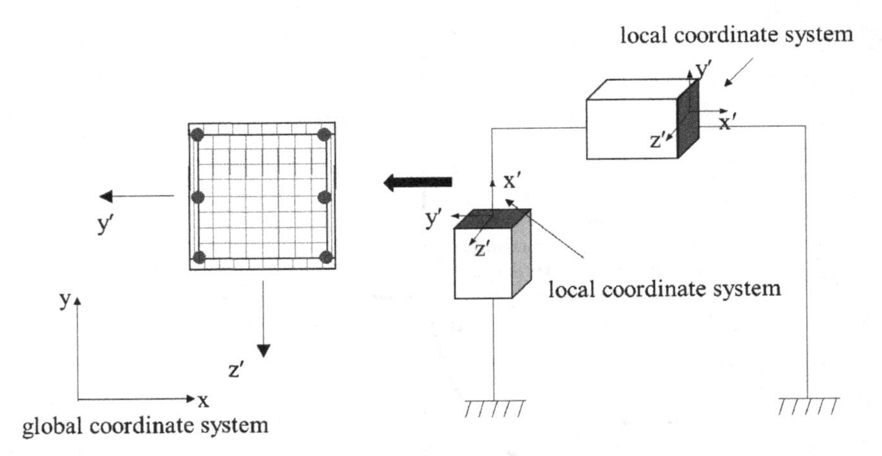

Fig. 4.15 The relationship of the global and local coordinate systems for the 2D nonlinear FRC frame.

> *Note*: The relationship between the global and local coordinate systems and the distribution of the steel bars are shown in Fig. 4.15. In the local coordinate system, the x' direction is always along the beam or column. So, for example, the command "element dispBeamColumn 1 1 3 5 1 1" indicates that the local x axis of element 1 (i.e., in the x' direction) is along the vertical upward direction from node 1 to node 3, while the z' direction is always outwardly perpendicular to the paper. The steel fiber and concrete fiber may have the same location, and their contributions are both accounted for (i.e., cumulated).

```
30 recorder Node -file output/disp_34.out -time
   -node 3 4 -dof 1 2 3 disp;
31 recorder Node -file output/reaction_12.out
   -time -node 1 2 -dof 1 2 3 reaction;
32 recorder Drift -file output/drift_1.out -time
   -iNode 1 2 -jNode 3 4 -dof 1 -perpDirn 2;
33 recorder Element -file output/force_12.out
   -time -ele 1 2 global- Force;
34 recorder Element -file output/foce_3.out
   -time -ele 3 globalForce;
35 recorder Element -file output/forcecolsec_1.out
   -time -ele 1 2 section 1 force;
```

```
36 recorder Element -file output/defocolsec_1.out
   -time -ele 1 2 section 1 deformation;
37 recorder Element -file output/forcecolsec_5.out
   -time -ele 1 2 section 5 force;
38 recorder Element -file output/defocolsec_5.out
   -time -ele 1 2 section 5 deformation;
39 recorder Element -file output/forcebeamsec_1.out
   -time -ele 3 section 1 force;
40 recorder Element -file output/defobeamsec_1.out
   -time -ele 3 section 1 deformation;
41 recorder Element -file output/forcebeamsec_5.out
   -time -ele 3 section 5 force;
42 recorder Element -file output/defobeamsec_5.out
   -time -ele 3 section 5 deformation;
```

Note: The command "recorder Element" is not consistent for different elements. Different elements may have different argument formats (e.g., "-ele 3 globalForce", "-ele 3 section 1 deformation"). In order to properly use recorder command for elements, users are recommended to check the member function "setResponse()" in the element's source code.

In this example, the source code of the 2D nonlinear beam element "dispBeamColumn2d" can be found on the website: http://opensees.berkeley.edu/→Developer→Browse the Source Code→trunk/SRC/element/dispBeamColumn/DispBeamColumn2d.cpp.
In the function "Response *DispBeamColumn2d::setResponse()" we can see the arguments stored in the variable "argv[0]" — "basicStiffness", "globalForce", and "localForce" — are all legal arguments.
We can also see from "DispBeamColumn2d :: setResponse ()" that the number corresponding to the argument "globalForce" is set to 1 (i.e., "theResponse = new ElementResponse (this, 1, P)"). The information to be recorded corresponding to "1" can be found in another function "DispBeamColumn2d :: getResponse ()":
if (responseID == 1) return eleInfo.setVector (this->get-ResistingForce ()).
So the resisting force in the global coordinate system will be recorded.

Note: To record the information of a Gauss point of this element (e.g., "-ele 3 section 1 deformation"), we can check the corresponding code in DispBeamColumn2d::setResponse(), in this case "else if (strstr(argv[0], "section") != 0)", "argv[1]" then is saved in a variable "sectionNum" denoting the Gauss point's number. The Gauss points are numbered from a location close to the first node to a location close to the second node of a frame element. The following code will then be executed: "theResponse = theSections [sectionNum-1] -> setResponse (&argv[2], argc-2, output);". That transfers the string after argv[2] (i.e., "deformation") to the corresponding section.

In this example, the 2D fiber section (from http://opensees.berkeley.edu/→Developer→Browse the Source Code→trunk/SRC/material/section/FiberSection2d.cpp) will handle the argument "deformation". In the FiberSection2d::setResponse(..) (which in fact calls its base class member function SectionForceDeformation::setResponse(..)), the number corresponding to the argument "deformation" is set to 1, and SectionForceDeformation::getResponse(..) records section deformation for the number 1. In this way, checking "setResponse()" and "getResponse()" of the corresponding element, section, material and so on yields the meaning of the arguments after "recorder Element".

```
43 pattern Plain 1 Linear {
44     eleLoad -ele 3 -type -beamUniform -65.33;
45 }
46 constraints Plain;
47 numberer Plain;
48 system BandGeneral;
49 test NormDispIncr 1.0e-8 6 2;
50 algorithm NewtonLineSearch 0.75;
51 integrator LoadControl 0.1;
52 analysis Static
53 analyze 10;
54 puts ''Gravity analysis done ...''
```

The above code performs the gravity analysis.

```
55 loadConst -time 0.0;
56 pattern Plain 2 Linear {
57     load 3 1.0 0.0 0.0;
58     load 4 1.0 0.0 0.0;
59 }
60 integrator DisplacementControl 3 1 0.001;
61 analyze 300;
62 puts "horizontal force analysis done ..."
```

The above code performs the static pushover analysis.

Analysis results: The pushover force on node 3 is shown in Fig. 4.16 as a function of the displacement. The output file "disp_34.out" records in column 1 the "time" (which is not actually time in the displacement control case, but rather the pushover force on node 3 or node 4) and in column 2 the horizontal displacement of node 3.

The nonlinear moment–curvature relationship for Section A of the RC column (Fig. 4.13) in the local coordinate system is shown in Fig. 4.17.

The curvature is recorded in column 3 of "defocolsec_1.out", while the moment is recorded in column 3 of the output file "forcecolsec_1.out".

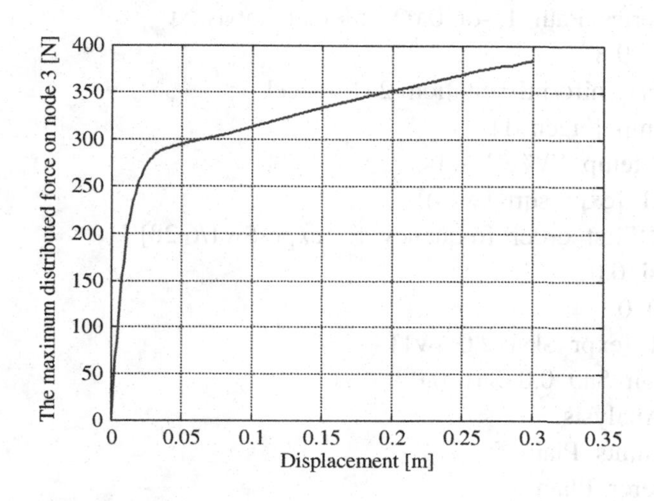

Fig. 4.16 The predicted pushover force vs. displacement.

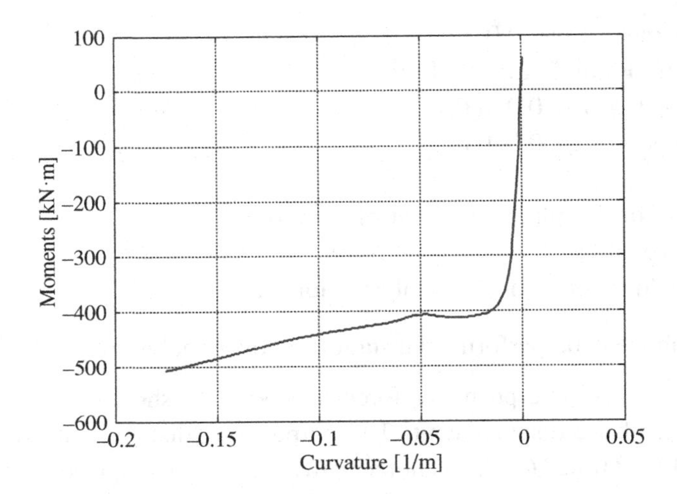

Fig. 4.17 Moment–curvature relationship for Section A.

Earthquake base excitation

In the seismic analysis, Lines 55–62 are replaced with the following code.

```
63 mass 3 20. 0. 0.
64 mass 4 20. 0. 0.
65 loadConst -time 0.0;
66 timeSeries Path 1 -dt 0.02 -filePath tabas.txt
   -factor 9.8;
67 pattern UniformExcitation 2 1 -accel 1;
68 set temp [eigen 1]
69 scan $temp ‘‘\%e’’ w1s
70 set w1 [expr sqrt($w1s)]
71 puts ‘‘First order frequency f: [expr $w1/6.28]’’
72 set ksi 0.02
73 set a0 0
74 set a1 [expr $ksi*2.0/$w1]
75 rayleigh $a0 0.0 $a1 0.0
76 wipeAnalysis
77 constraints Plain
78 numberer Plain
79 system BandGeneral
```

```
80 test NormDispIncr 1.0e-8 10 2
81 algorithm Newton
82 integrator Newmark 0.5 0.25
83 analysis Transient
84 analyze 1000 0.02
85 puts ''Ground motion analysis completed...''
```

Analysis results: The horizontal displacement history of the top node 3 is plotted in Fig. 4.18 using the data in the output file "disp_34.out" where the time and the horizontal displacement are recorded in the first and second columns, respectively.

The nonlinear moment–curvature responses in Section A of the FRC column (Fig. 4.13) in the local coordinate system are shown in Fig. 4.19. The curvatures and moments are recorded in column 3 of "defocolsec_1.out" and column 3 of "forcecolsec_1.out", respectively.

4.4 Example 4: Static and Dynamic Analyses of a 3D Inelastic Reinforced Concrete Frame

(A) Description of the example

The model is shown in Fig. 4.20. The cross-sections of the columns are nonlinear and the beams are elastic. The floors are rigid. Seismic

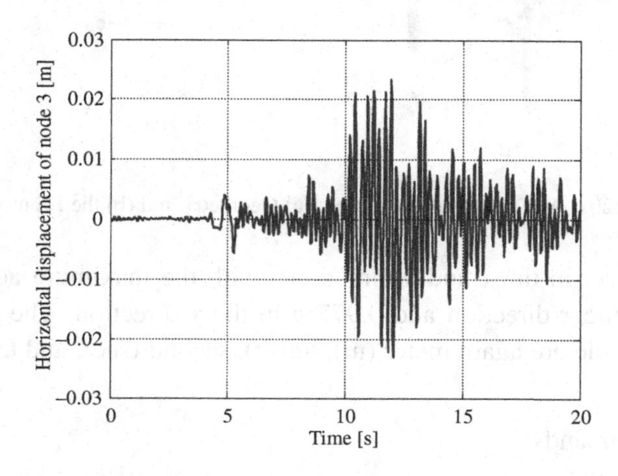

Fig. 4.18 Horizontal displacement history of the top node 3.

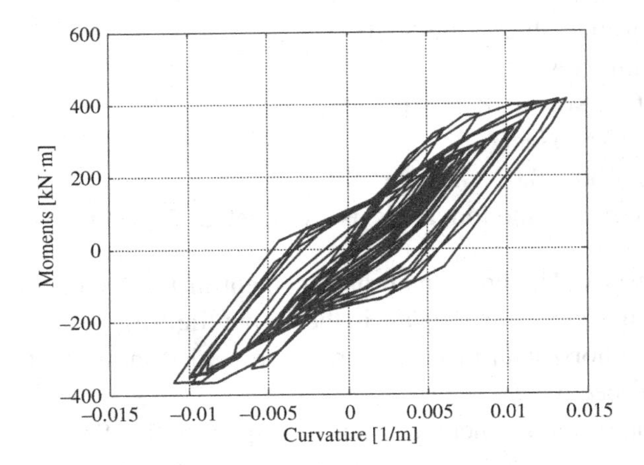

Fig. 4.19 Moment–curvature response of Section A.

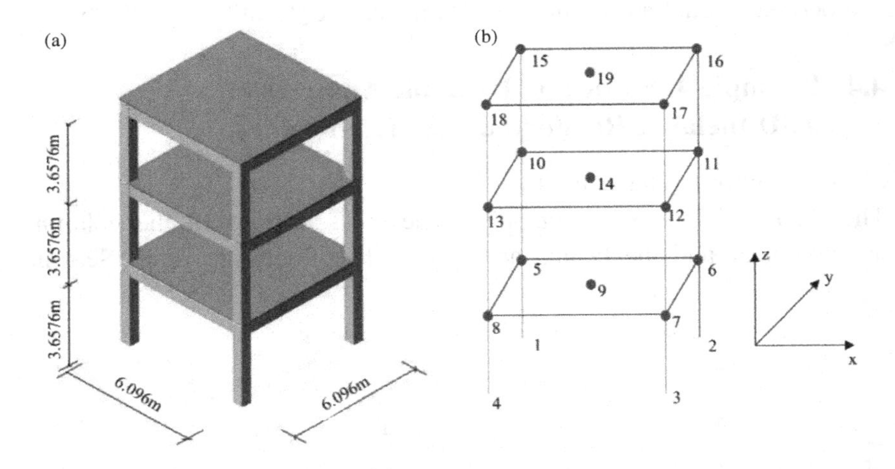

Fig. 4.20 A 3D inelastic RC frame: (a) Geometry and (b) the FE model.

excitations act in the x and y directions, with the maximum acceleration 0.900 g in the x direction and 0.977 g in the y direction. The units used in this example are again meter (m), ton (t), second (sec), and kiloNewton (kN).

(B) Tcl commands

```
1 model BasicBuilder -ndm 3 -ndf 6
```

In line 1, “3” indicates that the model is a 3D model, “ndf 6” indicates that each node has six degrees of freedom (3 translational and 3 rotational).

```
2 set h 3.6576
3 set by 6.096
4 set bx 6.096
```

Lines 2–4 define variables “h”, “bx”, and “by”, which denote the height of one floor, the length of the structure, and its width, respectively.

```
5 node 1  [expr -$bx/2] [expr $by/2] 0.0
6 node 2  [expr $bx/2]  [expr $by/2] 0.0
7 node 3  [expr $bx/2]  [expr -$by/2] 0.0
8 node 4  [expr -$bx/2] [expr -$by/2] 0.0
9 node 5  [expr -$bx/2] [expr $by/2] $h
10 node 6 [expr $bx/2] [expr $by/2] $h
11 node 7 [expr $bx/2] [expr -$by/2] $h
12 node 8 [expr -$bx/2] [expr -$by/2] $h
13 node 10 [expr -$bx/2] [expr $by/2] [expr 2*$h]
14 node 11 [expr $bx/2] [expr $by/2] [expr 2*$h]
15 node 12 [expr $bx/2] [expr -$by/2] [expr 2*$h]
16 node 13 [expr -$bx/2] [expr -$by/2] [expr 2*$h]
17 node 15 [expr -$bx/2] [expr $by/2] [expr 3*$h]
18 node 16 [expr $bx/2] [expr $by/2] [expr 3*$h]
19 node 17 [expr $bx/2] [expr -$by/2] [expr 3*$h]
20 node 18 [expr -$bx/2] [expr -$by/2] [expr 3*$h]
```

Lines 5–20 define the nodes of the frame elements.

```
21 node 9 0.0 0.0 $h
22 node 14 0.0 0.0 [expr 2*$h]
23 node 19 0.0 0.0 [expr 3*$h]
```

Lines 21–23 define 3 nodes located at the center of each rigid floor. These nodes are referred to as master nodes of each floor in the sense they are the center nodes of a rigid diaphragm defined later.

```
24 fix 1 1 1 1 1 1 1
25 fix 2 1 1 1 1 1 1
26 fix 3 1 1 1 1 1 1
27 fix 4 1 1 1 1 1 1
```

Lines 24–27 define the boundary condition constraints. For example, line 24 fixes the six degrees of freedom of node 1.

```
28 rigidDiaphragm 3 9 5 6 7 8
29 rigidDiaphragm 3 14 10 11 12 13
30 rigidDiaphragm 3 19 15 16 17 18
```

Lines 28–30 define three rigid floors (using Multi-Point Constraint). In line 28, the number "3" indicates that the rigid plane is perpendicular to the z direction, so the rigid floor is on the x-y plane. The number "9" is the floor's master node, and the numbers following the master node are tags of slave nodes (nodes 5, 6, 7, 8 for this floor). Line 28 thus creates a rigid floor, parallel to the x–y plane. Lines 29 and 30 define the other floors similarly.

In this example, the displacement of the rigid floor in the z direction and any rotation around the x and y axes are ignored for simplicity, so the following constraints are applied:

```
31 fix 9 0 0 1 1 1 0
32 fix 14 0 0 1 1 1 0
33 fix 19 0 0 1 1 1 0
```

Note: When a Multi-Point Constraint (MPC) such as "rigidDiaphragm" and "equalDOF" is used in a model, only "constraints Transformation" and "constraints Penalty" are allowed. "constraints Plain" works only for single-point constraints (e.g., "fix"), and not in MPC cases. When "Transformation" is used, any further constraint (such as "fix") on slave nodes will cause memory errors.

```
34 uniaxialMaterial Concrete01 1 -34473.8 -0.005
    -24131.66 -0.02
```

Line 34 defines the concrete material used in the core region. The material type is "Concrete01".

```
35 set fc 27579.04
36 uniaxialMaterial Concrete02 2 -{\$}fc -0.002
   0.0 -0.006
```

Lines 35 and 36 define the concrete material used in the cover. The material type is "Concrete02" (refer to http://opensees.berkeley.edu).

```
37 uniaxialMaterial Steel01 3 248200. 2.1e8 0.02
38 set E 24855585.89304
39 set GJ 68947600000000
40 uniaxialMaterial Elastic 10 {\$}GJ
```

Lines 37–40 define a elastic material in torsion.

This example defines a nonlinear fiber cross-section. The definition is the same as those in Section 4.3. For convenience, we can use a procedure defined in RCsection.tcl (which can be found using a web search) to build the fiber cross-section model.

```
41 source RCsection.tcl
```

Note that in this fiber section model the local coordinate system is in the *y*–*z* plane as shown in Fig. 4.21. The 3D local coordinate system will be explained later.

The arguments of RCsection are as follows:

id — the section number
h — section depth
b — section width
cover — cover depth (assumed uniform around the perimeter)
coreID — the number of the uniaxialMaterial assigned to each fiber in the core region
coverID — the number of the uniaxialMaterial assigned to each fiber in the cover region
steelID — the number of the uniaxialMaterial assigned to each reinforcing bar
numBars — the number of steel bars on each side of the section
barArea — each bar's cross-sectional area
nfCoreY — the number of fibers through the core depth in the *y* direction

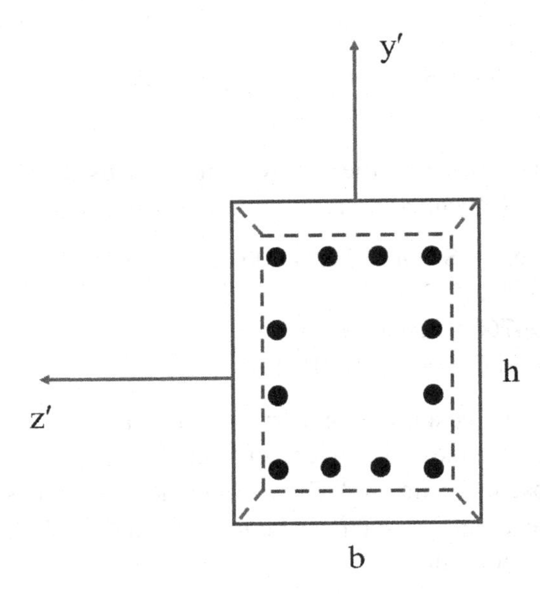

Fig. 4.21 The local coordinate system of the fiber section model.

nfCoreZ — the number of fibers through the core depth in the *z* direction
nfCoverY — the number of fibers through the cover depth in the *y* direction
nfCoverZ — the number of fibers through the cover depth in the *z* direction

In order to explain the procedure for using "RCsection", a letter "A" is added before the RCsection.tcl line numbers to distinguish them from the other Tcl codes, as follows:

```
A1 proc RCsection {id h b cover coreID coverID steelID numBars
   barArea nfCoreY nfCoreZ nfCoverY nfCoverZ} {
A2 set coverY [expr $h/2.0]
A3 set coverZ [expr $b/2.0]
A4 set ncoverY [expr -$coverY]
A5 set ncoverZ [expr -$coverZ]
A6 set coreY [expr $coverY-$cover]
A7 set coreZ [expr $coverZ-$cover]
```

```
A8 set ncoreY [expr -$coreY]
A9 set ncoreZ [expr -$coreZ]
```

A2–A9 define the depth of the concrete cover and the distances from the axis to the edges of the concrete core and cover regions.

```
A10 section fiberSec $id {
A11 patch quadr $coreID $nfCoreZ $nfCoreY $ncoreY $coreZ
    $ncoreY $ncoreZ $coreY $ncoreZ $coreY $coreZ
```

A11 defines the concrete core area. “quadr” indicates that the area is quadrilateral; “$coreID” is the previously defined material tag; and “$nfCoreY” and “$nfCoreZ” are the number of fiber elements divided along y' and z' directions in the core region. Subsequent parameters represent the coordinates of the four vertices in the core region, defined counterclockwise.

```
A12 patch quadr $coverID 1 $nfCoverY $ncoverY $coverZ $ncoreY
    $coreZ $coreY $coreZ $coverY $coverZ
A13 patch quadr $coverID 1 $nfCoverY $ncoreY $ncoreZ $ncoverY
    $ncoverZ $coverY $ncoverZ $coreY $ncoreZ
A14 patch quadr $coverID $nfCoverZ 1 $ncoverY $coverZ $ncoverY
    $ncoverZ $ncoreY $ncoreZ $ncoreY $coreZ
A15 patch quadr $coverID $nfCoverZ 1 $coreY $coreZ $coreY
    $ncoreZ $coverY $ncoverZ $coverY $coverZ
```

A13–A15 define the region of the concrete cover consisting of 4 quadrilaterals.

```
A16 layer straight $steelID $numBars $barArea $ncoreY $coreZ
    $ncoreY $ncoreZ
A17 layer straight $steelID $numBars $barArea $coreY $coreZ
    $coreY $ncoreZ
```

The command “layer straight” has been explained in Section 4.3. It defines the distinct fibers along a straight line, in this case the reinforcing steel bars along the local z' direction. The numbers following “layer straight” are the tag of the material, the number of bars along the line, the cross-sectional area of each bar, the y' and z' coordinates of the first bar, and the y' and z' coordinates of the last one.

```
A18 set spacingY [expr ($coreY-$ncoreY)/($numBars-1)]
A19 set numBars [expr $numBars-2]
A20 layer straight $steelID $numBars $barArea [expr $coreY-
    $spacingY] $coreZ [expr $ncoreY+$spacingY] $coreZ
A21 layer straight $steelID $numBars $barArea [expr $coreY-
    $spacingY] $ncoreZ [expr $ncoreY+$spacingY] $ncoreZ
        }
    }
```

Lines A18–A21 specify that the reinforcing bars are oriented along the local y' direction. That concludes the source code to define the procedure "RCsection". Now to establish a model.

```
42 RCsection 1 $d $d 0.04 1 2 3 3 0.00051 8 8 10 10
43 section Aggregator 2 10 T -section 1
```

After defining the flexural section (Section 1), in line 43, the command "section Aggregator" is used to identify that section with a torsion-resistant material (material #10).

```
44 set colSec 2
45 geomTransf Linear 1 1 0 0
46 set np 4
47 elemenet dispBeamColumn 1 1 5 $np $colSec 1
48 elemenet dispBeamColumn 2 2 6 $np $colSec 1
49 elemenet dispBeamColumn 3 3 7 $np $colSec 1
50 elemenet dispBeamColumn 4 4 8 $np $colSec 1
51 elemenet dispBeamColumn 5 5 10 $np $colSec 1
52 elemenet dispBeamColumn 6 6 11 $np $colSec 1
53 elemenet dispBeamColumn 7 7 12 $np $colSec 1
54 elemenet dispBeamColumn 8 8 13 $np $colSec 1
55 elemenet dispBeamColumn 9 10 15 $np $colSec 1
56 elemenet dispBeamColumn 10 11 16 $np $colSec 1
57 elemenet dispBeamColumn 11 12 17 $np $colSec 1
58 elemenet dispBeamColumn 12 13 18 $np $colSec 1
```

Lines 44–58 define elements. Line 45 defines the coordinate transformation of the 3D beam column elements, which is different from 2D coordinate transformation.

> *Note*: In OpenSees, a general 3D coordinate transformation is defined as follows:
> geomTransf Linear $transfNumber $vecxzX $vecxzY $vecxzZ
> "Linear" indicates that a linear transformation (not considering large displacement) will be used in the model. "$transfNumber" is the tag of the coordinate transformation. As with the two-dimensional beam model, the x' direction of the local coordinate system is along the rods (from first node to the second node of the element). For example, line 47 indicates that the x' direction is from node 1 to node 5.
>
> The last three parameters of "geomTransf" ("$vecxzX", "$vecxzY", and "$vecxzZ") define a new vector in the global coordinate system parallel to the $x'z'$ plane of the local coordinate system. The y' direction in the local coordinate system is thus the cross product of this new vector with the x' direction, and the local z' is the cross product of x' and y'.

Figure 4.22 shows a simple example of coordinate transformation. In the local coordinate system, the x' direction is defined as from node i to node j (along the rod). If the user defines a new direction (1, 1, 0) parallel to the $x'z'$ plane (as in the following line 64), the y' direction is the cross product of the new direction and x' direction, and z' is cross product of x' and y' (see Fig. 4.22).

```
59 setAbeam 0.278709
60 setIbeamzz 0.004315
61 setIbeamyy 0.002427
62 section Elastic 3 $E $Abeam $Ibeamzz $Ibeamyy $GJ 1.0
```

Lines 59–62 define an elastic section with number 3.

```
63 set beamSec 3
64 geomTransfLinear 2 1 1 0
65 set np 3
66 element dispBeamColumn 13 5 6 $np $beamSec 2
67 element dispBeamColumn 14 6 7 $np $beamSec 2
```

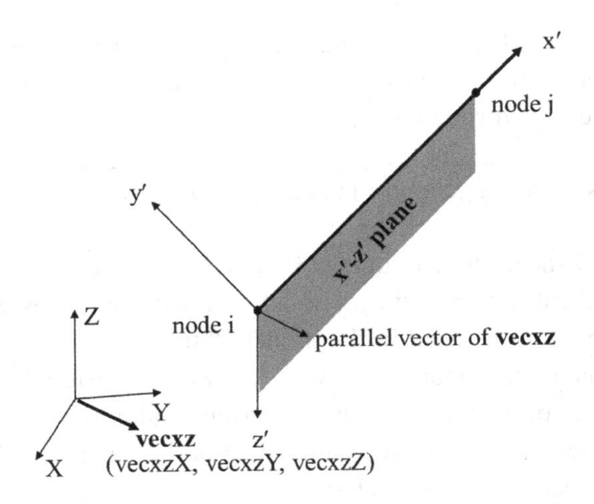

Fig. 4.22 An example of defining a local coordinate system.

```
68 element dispBeamColumn 15 7 8 $np $beamSec 2
69 element dispBeamColumn 16 8 7 $np $beamSec 2
70 element dispBeamColumn 17 10 11 $np $beamSec 2
71 element dispBeamColumn 18 11 12 $np $beamSec 2
72 element dispBeamColumn 19 12 13 $np $beamSec 2
73 element dispBeamColumn 20 13 10 $np $beamSec 2
74 element dispBeamColumn 21 15 16 $np $beamSec 2
75 element dispBeamColumn 22 16 17 $np $beamSec 2
76 element dispBeamColumn 23 17 18 $np $beamSec 2
77 element dispBeamColumn 24 18 15 $np $beamSec 2
```

Lines 63–77 define elements.

The local coordinate systems of element 1 (the left column) and element 13 (a beam) are shown in Fig. 4.23.

```
78 set g 9.8;
79 set m 30.0;
80 set i [expr $m*($bx*$bx+$by*$by)/12.0]
81 mass 9 $m $m 0 0 0 $i
82 mass 14 $m $m 0 0 0 $i
83 mass 19 $m $m 0 0 0 $i
```

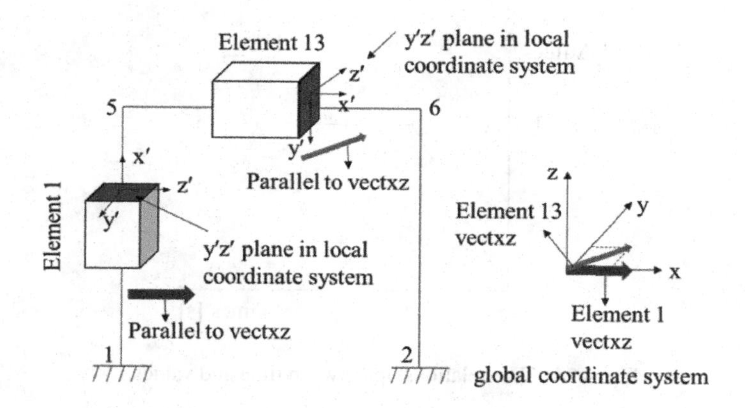

Fig. 4.23 The local coordinate systems of elements 1 and 13.

The mass is applied at each floor's master node. In a 3D system, the numbers following "mass" represent the number of the node, mass in the *x*, *y*, and *z* directions, and rotational inertia around the *x*, *y*, and *z* axes, respectively.

```
84 set p 74.0
85 pattern Plain 1 {Series --time {0.0 2.0 10000.0} --values {0.0
   1.0 1.0}} {foreach node {5 6 7 8 10 11 12 13 15 16 17 18} {load
   $node 0.0 0.0 -$p 0.0 0.0 0.0 }}
```

Lines 84 and 85 define a pattern, with the number 1 indicating a gravity load which increases linearly from 0.0 to 2.0 s, then stays constant (until some arbitrarily large time, say 10000 s). OpenSees keeps "time" in static analysis as it does in dynamic analysis. The "Series" command is used here to define the time series. The relationship between time and the values in the Series is shown in Fig. 4.24. At time 0 s, the nodes are subjected to zero force, while after time 2 s (until 10000 s) the nodes are subjected to a force in the *z* direction of magnitude -$p. (The absolute value of force is the values in "Series", i.e., 1.0 after 2 s, multiply by -$p, and the negative sign indicates that the force acts opposite to the specified positive direction.) At each time step OpenSees will automatically interpolate the external force and apply it. (For example, at 3.01 s the external force is -$p on both nodes 5 and 6.)

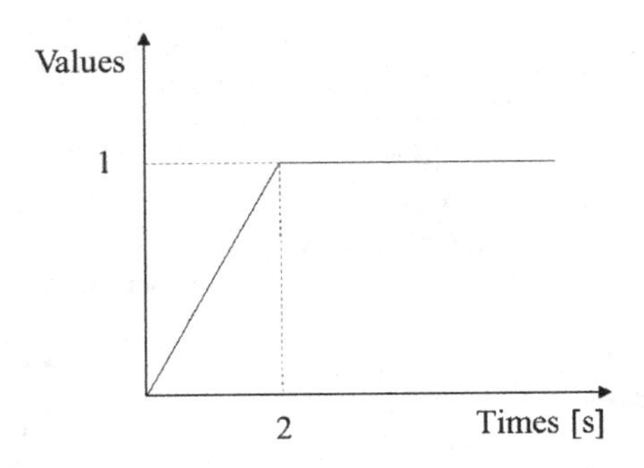

Fig. 4.24 The relationship between time and values.

> *Note*: After 10000 s, Fig. 4.24 suggests that the load suddenly drops to zero (even if the time is extremely close to 10000 s, e.g., 10000.00000000001 s). This may sometimes cause errors due to the inaccuracy of current time in OpenSees!

```
86 set tabasFN ''Path -filePath tabasFN.txt -dt 0.02 -factor $g''
87 set tabasFP ''Path -filePath tabasFP.txt -dt 0.02 -factor $g''
88 pattern UniformExcitation 2 1 -accel $tabasFN
89 pattern UniformExcitation 3 2 -accel $tabasFP
```

Lines 86–89 apply earthquake base excitations in two directions.

```
90 recorder Node -file node.out -time -node 9 14 19 -dof 1 2 3 4 5 6
   disp
```

Line 90 indicates that the deformation of nodes 9, 14, and 19 (the master nodes) in all six degrees of freedom will be recorded.

```
91 constraints Transformation
92 test EnergyIncr 1.0e-16 20 2
93 integrator LoadControl 1 1 1 1
94 algorithm Newton
95 system BandGeneral
96 numberer RCM
97 analysis Static
98 set startT [clock seconds]
99 analyze 3
```

Note: In this example, the gravity load no longer changes after 3.0 s (see line 85 and Fig. 4.24 for the definition of the pattern), so the command "loadConst -time 0" is not used.

Lines 91–99 are commands applying the (static) gravity analysis. Note that "constrains Plain" cannot be used in line 91 because it is a multi-point constraint. Line 98 defines a Tcl variable "startT" to record the start time of the FE analysis.

```
100 wipe Analysis
101 test EnergyIncr 1.0e-16 20 2
102 algorithm Newton
103 system BandGeneral
104 constraints Transformation
105 numberer RCM
106 integrator Newmark 0.55 0.275625
107 analysis Transient
108 analyze 2500 0.01
109 set endT [clock seconds]
110 puts ''finish time: [expr $endT-$startT] seconds.''
```

Lines 100–110 are Tcl commands implementing the seismic analysis.

A Newton iterative algorithm is used in this example to solve the nonlinear equations of motion at each time step. The Newmark method is used for the time integration, with the parameters α and β taken as 0.55 and 0.275625, respectively. Numerical damping will be applied, since α and β are not 0.5 and 0.25, respectively. (Slightly larger values of α and β may increase the numerical stability of FE analysis in some cases.) Line 109 records the finish time of the FE analysis and line 110 delivers the time used.

Note: The command "loadConst -time 0" is not used between the static and dynamic analyses, so the system time is 3.0 s at the start of the dynamic analysis. It is therefore necessary to add 3 s of unused acceleration records before the valid (used) ones in tabasFN.txt and tabasFP.txt. The time interval in those two files is 0.02 s, so 150 zeros have been added before the valid data.

A simpler alternative is to replace these elaborate operations by using the command "loadConst -time 0".

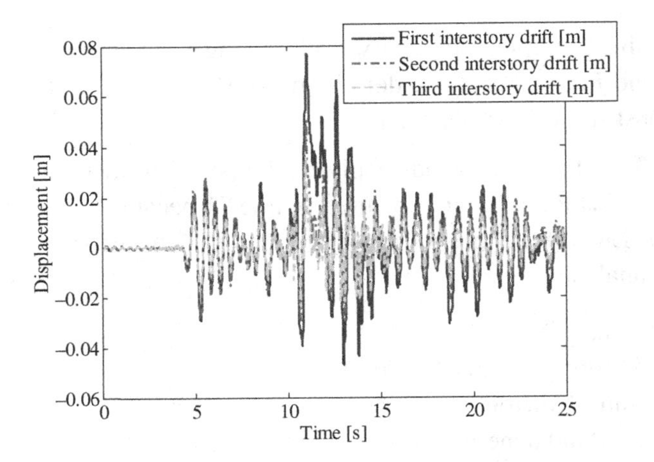

Fig. 4.25 Interstory drifts of the three stories in the x direction.

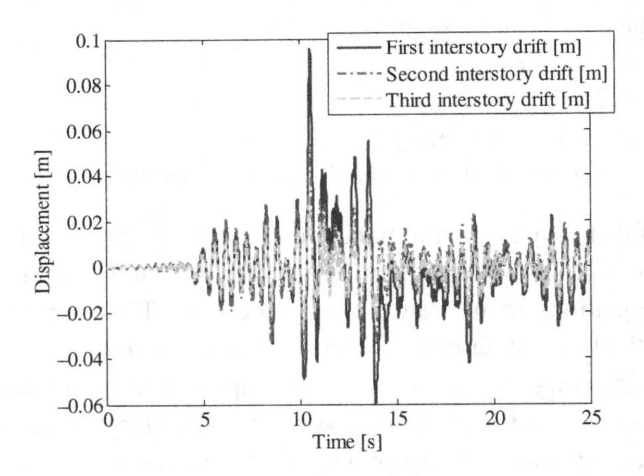

Fig. 4.26 Interstory drifts in the y direction.

Figure 4.25 displays the predicted interstory drifts in the x direction for the three stories. The maximum drifts are 0.0769 m, 0.0508 m, and 0.0355 m, occurring at 11.09 s, 11.03 s, and 10.99 s, respectively.

Figure 4.26 shows similar results for the y direction. The maximum values appear at 10.56 s, 10.50 s, and 10.47 s, with the corresponding magnitudes 0.0953 m, 0.0545 m, and 0.0281 m, respectively.

Chapter 5

Analyses for Reinforced Concrete Shear Walls

Shear wall is an important component in tall buildings for resisting wind or earthquake loads. This chapter introduces a multi-cross-line model (MCLM) that can provide a practically useful (i.e., relatively stable and accurate) simulation tool for nonlinear analysis of realistic high-rise buildings. Details may be found in Ref. [1]. The tcl example and source code of MCLM can be found in https://github.com/OpenSeesXMU.

Figure 5.1 shows an RC frame shear wall subjected to both vertical loading and horizontal pushover forces. The frame elements (see Fig. 5.1(b)) are modeled using nonlinear displacement-based Euler–Bernoulli elements and fiber sections with uniaxial concrete and steel material models (as shown in Figs. 5.1(d) and 5.1(e)). The shear wall is modeled using four-node quad element. The nonlinear behavior at each Gauss point is modeled using an MCLM (see Fig. 5.1(c)).

The Tcl commands are as follows:

```
1    wipe
2    model basic -ndm 2 -ndf 3
3    set L 1032;set H 3660;
4    set nL 6;set nH 10;
5    set nodeID 0;
6    set LColNodeID 1;
7    for {set i 0} {$i<=$nH} {incr i} {
8       set nodeID [expr $nodeID+1];
9       node $nodeID [expr 0*$deltL]
           [expr $i*$deltH];
10   }
```

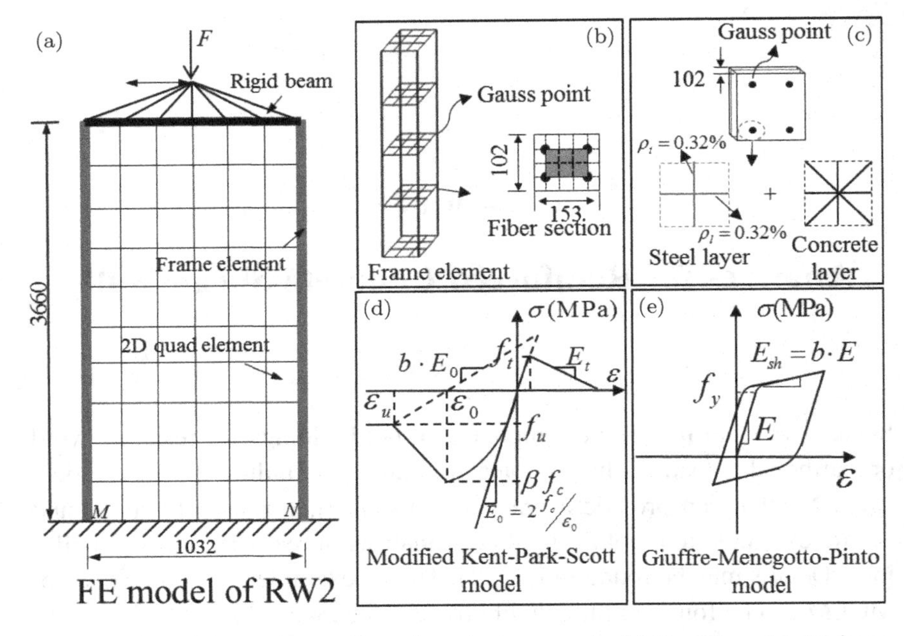

Fig. 5.1 The RC frame shear wall specimen in the example. (a) The FE model; (b) frame element and section properties; (c) quad element and section properties; (d) concrete constitutive model and (e) steel constitutive model.

```
11  set RColNodeID [expr $nodeID+1]
12  for {set i 0} {$i<=$nH} {incr i} {
13     set nodeID [expr $nodeID+1];
14     node $nodeID [expr $nL*$deltL][expr $i*$deltH];
15  }
16  set TopBeamNodeID [expr $nodeID+1]
17  for {set j 1} {$j<$nL} {incr j} {
18     set nodeID [expr $nodeID+1];
19     node $nodeID [expr $j*$deltL][expr $nH*$deltH];
20  }
21  fixY 0.0 1 1 1
22  set ControlNode [expr $nodeID+1]
23  set nodeID [expr $nodeID+1];
24  node $nodeID [expr $L/2.0] [expr $H+5/2.0];
```

```
25 uniaxialMaterial Concrete02 1 -47.6-0.006 -40.8 -0.028 0.1 2.6
   3.0e3;
26 uniaxialMaterial Concrete02 2 -42.8 -0.002 -8.2 -0.004 0.1 2.1 3.0e3;
27 uniaxialMaterial Steel02 3 395.2 200000.0 0.0185 18.0 0.925 0.15;
28 section Fiber 1 {
29    patch rect 1  20 10 -32 -57.5 32 57.5;
30    patch rect 2  20  1 -51 57.5 51 76.5;
31    patch rect 2  20  1 -51 -76.5 51 -57.5;
32    patch rect 2  1 10  -51 -57.5 -32 32;
33    patch rect 2  1 10  32 -57.5 51 57.5;
34    layer straight 3 4 71.2 -32 57.5 32 57.5;
35    layer straight 3 4 71.2 -32 -57.5 32 -57.5;
36 }
37 section Elastic 2 1e10 100 1e10;
38 geomTransf Linear 1
39 for {set i 1} {$i<=$nH} {incr i} {
40 element dispBeamColumn [expr 1000+$i] [expr
   $LColNodeID+$i-1] [expr $LColNodeID+$i] 5 1 1;
41  element dispBeamColumn [expr 2000+$i] [expr
   $RColNodeID+$i-1] [expr $RColNodeID+$i] 5 1 1;
42   }
```

Lines 39–42 define the left and right side columns.

```
43  element dispBeamColumn 3000 $ControlNode [expr
    $LColNodeID+$nH] 5 2 1
44  for {set i 1} {$i<$nL} {incr i} {
45  element dispBeamColumn [expr 3000+$i] $ControlNode [expr
    $TopBeamNodeID+$i-1] 5 2 1;
46  }
47  element dispBeamColumn [expr 3000+$nL] $ControlNode [expr
    $RColNodeID+$nH] 5 2 1;
```

Lines 43–47 define the rigid top beam. From line 48 on, the middle shear panel is simulated using 2D quad elements.

```
48   model basic -ndm 2 -ndf 2
49   set quadNodeID [expr $nodeID+1]
```

```
50    for {set i 0} {$i<=$nH} {incr i} {
51        for {set j 0} {$j<=$nL} {incr j} {
52            set nodeID [expr $nodeID+1];
53            node $nodeID [expr $j*$deltL]
              [expr $i*$deltH];
54        }
55    }
56    for {set i 1} {$i<=$nH} {incr i} {
57    equalDOF [expr $LColNodeID+$i] [expr
      $quadNodeID+($nL+1)*$i] 1 2;
58    equalDOF [expr $RColNodeID+$i] [expr
      $quadNodeID+($nL+1)*$i+$nL] 1 2;
59    }
60    for {set j 1} {$j<$nL} {incr j} {
61    equalDOF [expr $TopBeamNodeID +$j-1][expr $quadNodeID
      + ($nL+1) *$nH+$j] 1 2;
62    }
```

Lines 56–62 set the translational degrees of freedom of the node pairs (i.e., two nodes at the same location) in the rigid beam, the side columns and the quadrilateral elements being the same.

```
63    fixY 0.0 1 1;
64    set wfc 42.8;set wfy 336;set wE 2.0e5;
65    set rou1 2.4e-3; set rou2 2.4e-3;
66    uniaxialMaterial Steel02 11 336 2.0e5 0.035 18.0 0.925 0.15;
67    uniaxialMaterial SmearedConcrete 12 -42.8 -0.002
68    set pi 3.141592654;set t 102.0;
69    nDMaterial SmearedConcretePlaneStress 111 0.0 11 11 12 12 12
      12 [expr 0*$pi] [expr 0.5*$pi] $rou1 $rou2 $wfc $wfy $wE 0.002
```

SmearedConcretePlaneStress in line 69 defines a 2D plane-stressed RC material used to simulate the nonlinear behavior of the RC shear wall. The Steel02 (matID 11 in Line 66) and sMEAREDcONCRETE (matID 12 in

Line 67) are used to simulate the uniaxial steel and concrete bars used in the MCLM.

```
70    for {set i 1} {$i<=$nH} {incr i} {
71      for {set j 1} {$j<=$nL} {incr j} {
72        element quad [expr 5000+($i-1)*$nL+$j][expr $quadNodeID
          +($i-1)*($nL+1) + $j -1] [expr $quadNodeID + ($i-1)*($nL+1) +$j ]
          [expr $quadNodeID + ($i)*($nL+1) +$j ]  [expr
          $quadNodeID + ($i)*($nL+1) +$j -1] $t  PlaneStress 111;
73        }
74    }
75    pattern Plain 1 "Linear" {
76      load $ControlNode 0 -378000 0;
77    }
78    recorder Node -file disp.out -time -node
      $ControlNode -dof 1 disp
79    region 1 -nodeRange [expr $quadNodeID] [expr
      $quadNodeID+$nL] -node $LColNodeID $RColNodeID
80    recorder Node -file force.out -time -region
      1-dof 1 reaction;
81    system BandGeneral
82    constraints Transformation
83    numberer Plain
84    test NormDispIncr 1.0e-5 100 2
85    algorithm KrylovNewton
86    integrator LoadControl 0.1
87    analysis Static
88    analyze 10
89    loadConst -time 0.0
90    pattern Plain 2 "Linear" {
91      load $ControlNode 1 0 0;
92    }
93    system BandGeneral
94    constraints Transformation
95    numberer Plain
96    test NormDispIncr 1.0e-5 200 2
97    algorithm KrylovNewton
98    analysis Static
99    set iDmax "3.66 9.98 16.41 24.22 38.91 54.52
      71.05 71.28";
100   set Dincr 0.02;
101   set Fact 1.0;
```

```
102  source GeneratePeaks.tcl
103  set CycleType Full;

104  foreach Dmax $iDmax {
105    set iDstep [GeneratePeaks $Dmax $Dincr
       $CycleType $Fact];
106    for {set i 1} {$i <= 1} {incr i 1} {
107       set zeroD 0
108       set D0 0.0
109       foreach Dstep $iDstep {
110       set D1 $Dstep
111         set Dincr [expr $D1 - $D0]
112         integrator DisplacementControl $ControlNode 1 $Dincr
113         set ok [analyze 1];
114         set D0 $D1;# move to next step
115       }
116    }
117  }
```

Analysis results: Fig. 5.2 shows the shear force at the base vs. the hysteretic displacement response at the top. The predictions are compared with experimental results to evaluate the accuracy of the MCLM. The MCLM

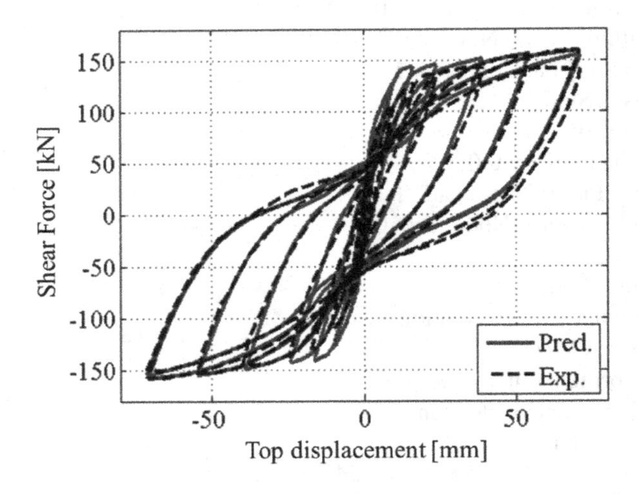

Fig. 5.2 Predicted vs. experimental responses.

can simulate the nonlinear behavior of a shear wall under cyclic loading with reasonable accuracy.

Reference

[1] Sun, B., Gu, Q., Zhang, P., *et al.* Efficient simulation of RC shear walls in high-rise buildings using a practical multi-cross-line-model. *J. Earthq. Eng.* 2021, *25*(9): 1732–1761.

Chapter 6

Soil–Structure Interaction (SSI) System

This chapter analyzes a simple 2D two-story frame structure on a soil foundation under earthquake loading. More complicated analyses of the soil–structure interaction (SSI) system may be conducted similarly.

(A) Description of the FE model

The SSI model is shown in Fig. 6.1. Nonlinear fiber sections are used in the beam–column elements. The cross-section of the middle column is $0.6\,\text{m} \times 0.5\,\text{m}$, while that of each side column is $0.5\,\text{m} \times 0.5\,\text{m}$ and that of the beams is $0.4\,\text{m} \times 0.4\,\text{m}$. The foundation consists of isolated RC footings below each column simulated as linear elastic models, while the soil is simulated using a multi-yield surface plasticity soil model.

Four different sets of material parameters are used in the four soil layers to represent the characteristics of the soil at different depths. Simple shear boundary conditions are employed to simulate the soil boundaries. The SSI system is subjected to an earthquake recorded in the 1940 El-Centro earthquake scaled by 3.0, with a peak ground acceleration (PGA) of $8.769\,\text{m/s}^2$.

Figure 6.1 shows the finite element discretization scheme representing the frame structure, the foundation, and the soil. For a detailed description, please refer to [1].

(B) Tcl commands

Step 1: Build the 2D superstructure frame model using Euler–Bernoulli frame elements with three degrees of freedom per node.

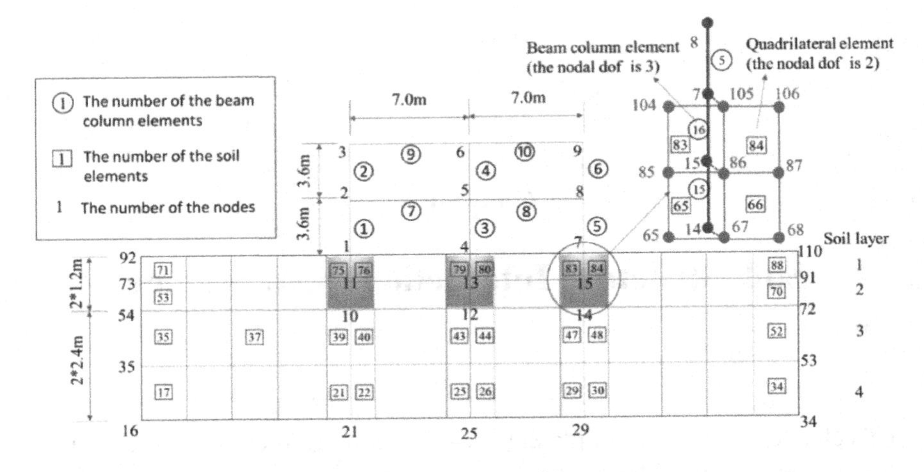

Fig. 6.1 The FE model of the SSI system.

```
1    wipe;
2    model BasicBuilder –ndm2 –ndf3
3    if {[file exists output] == 0} {
4      file mkdir output;
     }
5    set framemass1 15.0
6    set framemass2 30.0
7    set framemass3 4.0
8    node 1  0.0  0.0   -mass $framemass1 $framemass1 0.0
9    node 2  0.0  3.6   -mass $framemass1 $framemass1 0.0
10   node 3  0.0  7.2   -mass $framemass1 $framemass1 0.0
11   node 4  7.0  0.0   -mass $framemass2 $framemass2 0.0
12   node 5  7.0  3.6   -mass $framemass2 $framemass2 0.0
13   node 6  7.0  7.2   -mass $framemass2 $framemass2 0.0
14   node 7  14.0 0.0   -mass $framemass1 $framemass1 0.0
15   node 8  14.0 3.6   -mass $framemass1 $framemass1 0.0
16   node 9  14.0 7.2   -mass $framemass1 $framemass1 0.0
17   node 10 0.0 -2.4   -mass $framemass3 $framemass3 0.0
18   node 11 0.0 -1.2   -mass $framemass3 $framemass3 0.0
19   node 12 7.0 -2.4   -mass $framemass3 $framemass3 0.0
20   node 13 7.0 -1.2   -mass $framemass3 $framemass3 0.0
```

```
21  node 14 14.0 -2.4 -mass $framemass3 $framemass3 0.0
22  node 15 14.0 -1.2 -mass $framemass3 $framemass3 0.0
```

Lines 5–22 define nodes with lumped mass.

```
23  recorder Node -file output/disp6.out -time -node 6 -dof 1 2 disp
24  recorder Node -file output/disp5.out -time -node 5 -dof 1 2 disp
25  recorder Node -file output/disp4.out -time -node 4 -dof 1 2 disp
26  set upperload1 [expr -$framemass1*10.0]
27  set upperload2 [expr -$framemass2*10.0]
28  set download3 [expr -$framemass3*10.0]
29  pattern Plain 2 "Constant" {
30     load 1 0.0 $upperload1 0.0
31     load 2 0.0 $upperload1 0.0
32     load 3 0.0 $upperload1 0.0
33     load 4 0.0 $upperload2 0.0
34     load 5 0.0 $upperload2 0.0
35     load 6 0.0 $upperload2 0.0
36     load 7 0.0 $upperload1 0.0
37     load 8 0.0 $upperload1 0.0
38     load 9 0.0 $upperload1 0.0
39     load 10 0.0 $download3 0.0
40     load 11 0.0 $download3 0.0
41     load 12 0.0 $download3 0.0
42     load 13 0.0 $download3 0.0
43     load 14 0.0 $download3 0.0
44     load 15 0.0 $download3 0.0
45  }
```

Lines 29–45 define a pattern to apply the gravity load.

```
46  uniaxialMaterial Concrete01 1 -27588.5  -0.002 0.0 -0.008
47  uniaxialMaterial Concrete01 2 -34485.6  -0.004 -20691.4 -0.014
48  uniaxialMaterial Hardening 3 2.0e8 248200.0 0.0 1.6129e6
49  uniaxialMaterial Concrete01 4 -27588.5  -0.002 0.0 -0.008
50  uniaxialMaterial Concrete01 5 -34485.6  -0.004 -20691.4 -0.014
51  uniaxialMaterial Hardening 6 2.0e8 248200.0 0.0 1.6129e6
```

Lines 46–51 define the uniaxial concrete and steel materials. Lines 48 and 51 define the steel bars in the RC columns. The constitutive behavior of the reinforcing steel is modeled using one-dimensional (1D) J2 plasticity with kinematic and isotropic linear hardening. The numbers following "Hardening" are the tag of the material, the elastic modulus of the steel bars, yield strength, isotropic hardening modulus, and the kinematic hardening modulus.

```
52  section Fiber 1 {
53     patch quad 2 1 12 -0.25 0.20 -0.25 -0.20 0.25-0.20 0.25 0.20
54     patch quad 1 1 14 -0.30 -0.20 -0.30 -0.25 0.30-0.25 0.30 -0.20
55     patch quad 1 1 14 -0.30 0.25 -0.30 0.20 0.30 0.20 0.30 0.25
56     patch quad 1 1 2 -0.30 0.20 -0.30 -0.20 -0.25 -0.20 -0.25 0.20
57     patch quad 1 1 2 0.25 0.20 0.25 -0.20 0.30 -0.20 0.30 0.20
58     layer straight 3 3 0.000645 -0.20 0.20 -0.20 -0.20
59     layer straight 3 3 0.000645 0.20 0.20 0.20 -0.20
60  }
61  section Fiber 2 {
62     patch quad 2 1 10 -0.20 0.20 -0.20 -0.20 0.20 -0.20 0.20 0.20
63     patch quad 1 1 12 -0.25 -0.20 -0.25 -0.25 0.25 -0.25 0.25 -0.20
64     patch quad 1 1 12 -0.25 0.25 -0.25 0.20 0.25 0.20 0.25 0.25
65     patch quad 1 1 2 -0.25 0.20 -0.25 -0.20 -0.20 -0.20 -0.20 0.20
66     patch quad 1 1 2 0.20 0.20 0.20 -0.20 0.25 -0.20 0.25 0.20
67     layer straight 3 3 0.00051 -0.20 0.20 -0.20 -0.20
68     layer straight 3 3 0.00051 0.20 0.20 0.20 -0.20}
69  section Fiber 3 {
70     patch quad 1 1 12 -0.25 0.20 -0.25 -0.20 0.25 -0.20 0.25 0.20
71     layer straight 3 2 0.000645 -0.20 0.20 -0.20 -0.20
72     layer straight 3 2 0.000645 0.20 0.20 0.20 -0.20}
73  section Fiber 4 {
74     patch quad 5 1 12 -0.25 0.20 -0.25 -0.20 0.25 -0.20 0.25 0.20
75     patch quad 4 1 14 -0.30 -0.20 -0.30 -0.25 0.30 -0.25 0.30 -0.20
76     patch quad 4 1 14 -0.30 0.25 -0.30 0.20 0.30 0.20 0.30 0.25
77     patch quad 4 1 2 -0.30 0.20 -0.30 -0.20 -0.25 -0.20 -0.25 0.20
78     patch quad 4 1 2 0.25 0.20 0.25 -0.20 0.30 -0.20 0.30 0.20
79     layer straight 6 3 0.000645 -0.20 0.20 -0.20 -0.20
80     layer straight 6 3 0.000645 0.20 0.20 0.20 -0.20
81  }
82  section Fiber 5 {
83     patch quad 5 1 10 -0.20 0.20 -0.20 -0.20 0.20-0.20 0.20 0.20
84     patch quad 4 1 12 -0.25 -0.20 -0.25 -0.25 0.25 -0.25 0.25 -0.20
```

```
85    patch quad 4 1 12 -0.25 0.25 -0.25 0.20 0.250.20 0.25 0.25
86    patch quad 4 1 2 -0.25 0.20 -0.25 -0.20 -0.20-0.20 -0.20 0.20
87    patch quad 4 1 2 0.20 0.20 0.20 -0.20 0.25-0.20 0.25 0.20
88    layer straight 6 3 0.00051 -0.20 0.20 -0.20 -0.20
89    layer straight 6 3 0.00051 0.20 0.20 0.20 -0.20
}
```

Lines 52–89 define the RC fiber sections. The first three sections are used in the superstructure and last two, for the foundation columns. Sections 1, 2, and 3 describe the middle columns, side columns, and beams of the superstructure, respectively, while Sections 4 and 5 represent the side and middle columns of the foundation, respectively.

```
90  set nP 4
91  geomTransf Linear 1
92  element dispBeamColumn 1 1 2 $nP 2 1
93  element dispBeamColumn 2 2 3 $nP 2 1
94  element dispBeamColumn 3 4 5 $nP 1 1
95  element dispBeamColumn 4 5 6 $nP 1 1
96  element dispBeamColumn 5 7 8 $nP 2 1
97  element dispBeamColumn 6 8 9 $nP 2 1
98  element dispBeamColumn 7 2 5 $nP 3 1
99  element dispBeamColumn 8 5 8 $nP 3 1
100 element dispBeamColumn 9 3 6 $nP 3 1
101 element dispBeamColumn 10 6 9 $nP 3 1
102 element dispBeamColumn 11 10 11 $nP 4 1
103 element dispBeamColumn 12 11 1 $nP 4 1
104 element dispBeamColumn 13 12 13 $nP 5 1
105 element dispBeamColumn 14 13 4 $nP 5 1
106 element dispBeamColumn 15 14 15 $nP 4 1
107 element dispBeamColumn 16 15 7 $nP 4 1
```

Lines 92–97 define the frame elements for the columns of the superstructure. Lines 98–101 define the frame elements for the superstructure's beams. Lines 102–107 define frame elements of the substructure in the foundation. The left foot foundation consists of the substructure frame elements #11 and #12 and the soil elements #57,

#58, #75, #76, as shown in Fig. 6.1. The middle and right foot foundations are similar to the left one.

```
108 recorder Element -ele 1 2 -file output/Deformation12.out -time
    section 2 deformations
109 recorder Element -ele 1 2 -file output/Force12.out -time section 2
    force
110 recorder Element -ele 3 4 -file output/Deformation34.out -time
    section 2 deformations
111 recorder Element -ele 3 4 -file output/Force34.out -time section 2
    force
112 recorder Element -ele 7 9 -file output/Deformation79.out -time
    section 3 deformations
113 recorder Element -ele 7 9 -file output/Force79.out -time section 3
    force
114 recorder Element -ele 7 -time -file output/steelstress7.out section
    3 fiber -0.2286 0.2286 stress
115 recorder Element -ele 7 -time -file output/steelstrain7.out section
    3 fiber -0.2286 0.2286 strain
116 recorder Element -ele 7 -time -file output/concretestress7.out
    section 3 fiber 0.0 0.0 stress
117 recorder Element -ele 7 -time -file output/concretestrain7.out
    section 3 fiber 0.0 0.0 strain
```

Note: In the 2D SSI problem, soil node has two degrees of freedom (dof) while the frame elements in both the super- and sub-structures have 3 dof. The SSI model thus consists of two different models (as indicated by using the command "model" again in the line 119): a soil model and a structure model with different dof. The two models are connected by "tying" together the corresponding node pairs (i.e., a node of the structure model and a node of the soil model at the same location, e.g., nodes 7 and 105), thus constraining each node pair to have the same translational displacements. This is a simplified method to consider the connection between pile and soil.

Step 2: Build the 2D soil model with two degrees of freedom per node.
The soil model is built as follows:

```
118 set g -19.6
119 model basic -ndm 2 -ndf 2
120 node 16 -9.2 -7.2
121 node 17 -7.2 -7.2
122 node 18 -5.2 -7.2
123 node 19 -3.2 -7.2
124 node 20 -1.2 -7.2
125 node 21 0.0 -7.2
126 node 22 1.2 -7.2
127 node 23 3.5 -7.2
128 node 24 5.8 -7.2
129 node 25 7.0 -7.2
130 node 26 8.2 -7.2
131 node 27 10.5 -7.2
132 node 28 12.8 -7.2
133 node 29 14.0 -7.2
134 node 30 15.2 -7.2
135 node 31 17.2 -7.2
136 node 32 19.2 -7.2
137 node 33 21.2 -7.2
138 node 34 23.2 -7.2
139 node 35 -9.2 -4.8
140 node 36 -7.2 -4.8
141 node 37 -5.2 -4.8
142 node 38 -3.2 -4.8
143 node 39 -1.2 -4.8
144 node 40 0.0 -4.8
145 node 41 1.2 -4.8
146 node 42 3.5 -4.8
147 node 43 5.8 -4.8
148 node 44 7.0 -4.8
149 node 45 8.2 -4.8
```

```
150 node 46 10.5 -4.8
151 node 47 12.8 -4.8
152 node 48 14.0 -4.8
153 node 49 15.2 -4.8
154 node 50 17.2 -4.8
155 node 51 19.2 -4.8
156 node 52 21.2 -4.8
157 node 53 23.2 -4.8
158 node 54 -9.2 -2.4
159 node 55 -7.2 -2.4
160 node 56 -5.2 -2.4
161 node 57 -3.2 -2.4
162 node 58 -1.2 -2.4
163 node 59 0.0 -2.4
164 node 60 1.2 -2.4
165 node 61 3.5 -2.4
166 node 62 5.8 -2.4
167 node 63 7.0 -2.4
168 node 64 8.2 -2.4
169 node 65 10.5 -2.4
170 node 66 12.8 -2.4
171 node 67 14.0 -2.4
172 node 68 15.2 -2.4
173 node 69 17.2 -2.4
174 node 70 19.2 -2.4
175 node 71 21.2 -2.4
176 node 72 23.2 -2.4
177 node 73 -9.2 -1.2
178 node 74 -7.2 -1.2
179 node 75 -5.2 -1.2
180 node 76 -3.2 -1.2
181 node 77 -1.2 -1.2
182 node 78 0.0 -1.2
183 node 79 1.2 -1.2
184 node 80 3.5 -1.2
185 node 81 5.8 -1.2
```

```
186 node 82 7.0 -1.2
187 node 83 8.2 -1.2
188 node 84 10.5 -1.2
189 node 85 12.8 -1.2
190 node 86 14.0 -1.2
191 node 87 15.2 -1.2
192 node 88 17.2 -1.2
193 node 89 19.2 -1.2
194 node 90 21.2 -1.2
195 node 91 23.2 -1.2
196 node 92 -9.2 0.0
197 node 93 -7.2 0.0
198 node 94 -5.2 0.0
199 node 95 -3.2 0.0
200 node 96 -1.2 0.0
201 node 97 0.0 0.0
202 node 98 1.2 0.0
203 node 99 3.5 0.0
204 node 100 5.8 0.0
205 node 101 7.0 0.0
206 node 102 8.2 0.0
207 node 103 10.5 0.0
208 node 104 12.8 0.0
209 node 105 14.0 0.0
210 node 106 15.2 0.0
211 node 107 17.2 0.0
212 node 108 19.2 0.0
213 node 109 21.2 0.0
214 node 110 23.2 0.0
```

Lines 120–214 define the nodes in the soil.

```
215 fix 16 1 1
216 fix 17 1 1
217 fix 18 1 1
218 fix 19 1 1
219 fix 20 1 1
```

```
220 fix 21 1 1
221 fix 22 1 1
222 fix 23 1 1
223 fix 24 1 1
224 fix 25 1 1
225 fix 26 1 1
226 fix 27 1 1
227 fix 28 1 1
228 fix 29 1 1
229 fix 30 1 1
230 fix 31 1 1
231 fix 32 1 1
232 fix 33 1 1
233 fix 34 1 1
```

Lines 215–233 define the soil boundary conditions.

```
234 nDMaterial MultiYieldSurfaceClay 101 2 2.0 54450 1.6e5
    33.0 0.1
235 nDMaterial MultiYieldSurfaceClay 102 2 2.0 33800 1.0e5
    26.0 0.1
236 nDMaterial MultiYieldSurfaceClay 103 2 2.0  61250 1.8e5
    35.0 0.1
237 nDMaterial MultiYieldSurfaceClay 104 2 2.0 96800 2.9e5
    44.0 0.1
238 nDMaterial MultiYieldSurfaceClay 100 2 2.0 2e7 1.0e6
    21000.0 50.0
```

Lines 234–237 model a multi-yield surface soil material. The numbers following "MultiYieldSurfaceClay" represent the tag of the material, the dimension, the mass density of the soil (in this example, the material density is 0, while element density is not 0), the shear modulus, the bulk modulus, the yield strength, and its maximum shear strain. The four material definitions represent four different soil layers.

Line 238 defines the material of the foundation. Although plastic model "MultiYieldSurfaceClay" is used, the yield strength is set to be large enough such that the foundations remain elastic under the loads of this example.

```
239 element quad 17 16 17 36 35 0.60 "PlaneStrain" 104 0 0.0 0 $g
240 element quad 18 17 18 37 36 0.60 "PlaneStrain" 104 0 0.0 0 $g
241 element quad 19 18 19 38 37 0.60 "PlaneStrain" 104 0 0.0 0 $g
242 element quad 20 19 20 39 38 0.60 "PlaneStrain" 104 0 0.0 0 $g
243 element quad 21 20 21 40 39 0.60 "PlaneStrain" 104 0 0.0 0 $g
244 element quad 22 21 22 41 40 0.60 "PlaneStrain" 104 0 0.0 0 $g
245 element quad 23 22 23 42 41 0.60 "PlaneStrain" 104 0 0.0 0 $g
246 element quad 24 23 24 43 42 0.60 "PlaneStrain" 104 0 0.0 0 $g
247 element quad 25 24 25 44 43 0.60 "PlaneStrain" 104 0 0.0 0 $g
248 element quad 26 25 26 45 44 0.60 "PlaneStrain" 104 0 0.0 0 $g
249 element quad 27 26 27 46 45 0.60 "PlaneStrain" 104 0 0.0 0 $g
250 element quad 28 27 28 47 46 0.60 "PlaneStrain" 104 0 0.0 0 $g
251 element quad 29 28 29 48 47 0.60 "PlaneStrain" 104 0 0.0 0 $g
252 element quad 30 29 30 49 48 0.60 "PlaneStrain" 104 0 0.0 0 $g
253 element quad 31 30 31 50 49 0.60 "PlaneStrain" 104 0 0.0 0 $g
254 element quad 32 31 32 51 50 0.60 "PlaneStrain" 104 0 0.0 0 $g
255 element quad 33 32 33 52 51 0.60 "PlaneStrain" 104 0 0.0 0 $g
256 element quad 34 33 34 53 52 0.60 "PlaneStrain" 104 0 0.0 0 $g
257 element quad 35 35 36 55 54 0.60 "PlaneStrain" 103 0 0.0 0 $g
258 element quad 36 36 37 56 55 0.60 "PlaneStrain" 103 0 0.0 0 $g
259 element quad 37 37 38 57 56 0.60 "PlaneStrain" 103 0 0.0 0 $g
260 element quad 38 38 39 58 57 0.60 "PlaneStrain" 103 0 0.0 0 $g
261 element quad 39 39 40 59 58 0.60 "PlaneStrain" 103 0 0.0 0 $g
262 element quad 40 40 41 60 59 0.60 "PlaneStrain" 103 0 0.0 0 $g
263 element quad 41 41 42 61 60 0.60 "PlaneStrain" 103 0 0.0 0 $g
264 element quad 42 42 43 62 61 0.60 "PlaneStrain" 103 0 0.0 0 $g
265 element quad 43 43 44 63 62 0.60 "PlaneStrain" 103 0 0.0 0 $g
266 element quad 44 44 45 64 63 0.60 "PlaneStrain" 103 0 0.0 0 $g
267 element quad 45 45 46 65 64 0.60 "PlaneStrain" 103 0 0.0 0 $g
268 element quad 46 46 47 66 65 0.60 "PlaneStrain" 103 0 0.0 0 $g
269 element quad 47 47 48 67 66 0.60 "PlaneStrain" 103 0 0.0 0 $g
270 element quad 48 48 49 68 67 0.60 "PlaneStrain" 103 0 0.0 0 $g
271 element quad 49 49 50 69 68 0.60 "PlaneStrain" 103 0 0.0 0 $g
```

```
272 element quad 50 50 51 70 69 0.60 "PlaneStrain" 103 0 0.0 0 $g
273 element quad 51 51 52 71 70 0.60 "PlaneStrain" 103 0 0.0 0 $g
274 element quad 52 52 53 72 71 0.60 "PlaneStrain" 103 0 0.0 0 $g
275 element quad 53 54 55 74 73 0.60 "PlaneStrain" 102 0 0.0 0 $g
276 element quad 54 55 56 75 74 0.60 "PlaneStrain" 102 0 0.0 0 $g
277 element quad 55 56 57 76 75 0.60 "PlaneStrain" 102 0 0.0 0 $g
278 element quad 56 57 58 77 76 0.60 "PlaneStrain" 102 0 0.0 0 $g
279 element quad 57 58 59 78 77 0.60 "PlaneStrain" 100 0 0.0 0 $g
280 element quad 58 59 60 79 78 0.60 "PlaneStrain" 100 0 0.0 0 $g
281 element quad 59 60 61 80 79 0.60 "PlaneStrain" 102 0 0.0 0 $g
282 element quad 60 61 62 81 80 0.60 "PlaneStrain" 102 0 0.0 0 $g
283 element quad 61 62 63 82 81 0.60 "PlaneStrain" 100 0 0.0 0 $g
284 element quad 62 63 64 83 82 0.60 "PlaneStrain" 100 0 0.0 0 $g
285 element quad 63 64 65 84 83 0.60 "PlaneStrain" 102 0 0.0 0 $g
286 element quad 64 65 66 85 84 0.60 "PlaneStrain" 102 0 0.0 0 $g
287 element quad 65 66 67 86 85 0.60 "PlaneStrain" 100 0 0.0 0 $g
288 element quad 66 67 68 87 86 0.60 "PlaneStrain" 100 0 0.0 0 $g
289 element quad 67 68 69 88 87 0.60 "PlaneStrain" 102 0 0.0 0 $g
290 element quad 68 69 70 89 88 0.60 "PlaneStrain" 102 0 0.0 0 $g
291 element quad 69 70 71 90 89 0.60 "PlaneStrain" 102 0 0.0 0 $g
292 element quad 70 71 72 91 90 0.60 "PlaneStrain" 102 0 0.0 0 $g
293 element quad 71 73 74 93 92 0.60 "PlaneStrain" 101 0 0.0 0 $g
294 element quad 72 74 75 94 93 0.60 "PlaneStrain" 101 0 0.0 0 $g
295 element quad 73 75 76 95 94 0.60 "PlaneStrain" 101 0 0.0 0 $g
296 element quad 74 76 77 96 95 0.60 "PlaneStrain" 101 0 0.0 0 $g
297 element quad 75 77 78 97 96 0.60 "PlaneStrain" 100 0 0.0 0 $g
298 element quad 76 78 79 98 97 0.60 "PlaneStrain" 100 0 0.0 0 $g
299 element quad 77 79 80 99 98 0.60 "PlaneStrain" 101 0 0.0 0 $g
300 element quad 78 80 81 100 99 0.60 "PlaneStrain" 101 0 0.0 0 $g
301 element quad 79 81 82 101 100 0.60 "PlaneStrain" 100 0 0.0 0 $g
302 element quad 80 82 83 102 101 0.60 "PlaneStrain" 100 0 0.0 0 $g
303 element quad 81 83 84 103 102 0.60 "PlaneStrain" 101 0 0.0 0 $g
304 element quad 82 84 85 104 103 0.60 "PlaneStrain" 101 0 0.0 0 $g
305 element quad 83 85 86 105 104 0.60 "PlaneStrain" 100 0 0.0 0 $g
306 element quad 84 86 87 106 105 0.60 "PlaneStrain" 100 0 0.0 0 $g
307 element quad 85 87 88 107 106 0.60 "PlaneStrain" 101 0 0.0 0 $g
```

```
308 element quad 86 88 89 108 107 0.60 "PlaneStrain" 101 0 0.0 0 $g
309 element quad 87 89 90 109 108 0.60 "PlaneStrain" 101 0 0.0 0 $g
310 element quad 88 90 91 110 109 0.60 "PlaneStrain" 101 0 0.0 0 $g
```

Lines 239–310 define the four-node quadrilateral (or quad) soil elements. The numbers following "quad" are the tag of the element, its four nodes (counterclockwise), and the thickness of the element. "PlaneStrain" indicates that the element is a plain strain element, and the numbers following "PlaneStrain" are the tag of the material, the surface pressure, the element's mass density, and the constant body forces in the x and y directions.

The foundation consists of the substructure frame elements and soil elements #57, #58, #75, #76, #61, #62, #79, #80, #65, #66, #83, and #84. The material number for the soil elements is 100. Lines 239–256 define the elements of the 4th layer of the soil, lines 257–274 define the 3rd layer, lines 275–292 define the second layer, and lines 293–310 define the first layer (refer to Fig. 6.1).

> *Note*: The "material" command allows defining a material density and the "element" command allows defining an element density for the quad elements of the soil. It is recommended to define the material density only (setting the element density to 0.0) to avoid possible errors. In this example, the density of a quad element is assumed to be 0.0 t/m^3, and the density of the multi-yield surface material is set at 2.0 t/m^3. The actual density is 2.0 t/m^3.

Step 3: Set up the boundary conditions.

```
311 # equalDOF 16 34 1 2 ‘‘#’’ means this line is commented!
312 equalDOF 35 53 1 2
313 equalDOF 54 72 1 2
314 equalDOF 73 91 1 2
315 equalDOF 92 110 1 2
```

To simulate simple shear, the displacements of any two nodes at the same depth on the left and right boundaries must be the same. Note that line 311 must be deleted, because nodes 16 and 34 are both fixed, and further constraint on those nodes may cause a memory error.

```
316 equalDOF  1  97 1 2
317 equalDOF 11  78 1 2
318 equalDOF 10  59 1 2
319 equalDOF 4  101 1 2
320 equalDOF 13  82 1 2
321 equalDOF 12  63 1 2
322 equalDOF 7  105 1 2
323 equalDOF 15  86 1 2
324 equalDOF 14  67 1 2
```

Lines 316–324 connect the frame structure model with the soil model, ensuring that the translational displacements of each node pair are the same.

This is a simple way to model SSI systems, but if it is necessary to consider sliding between the piles and the soil, the diameter of piles and/or the no-tension capacity of the soil, more complicated simulation techniques (e.g., using contact elements) may be used to get realistic SSI behavior.

```
325 foreach theNode { 6 5 4 13 12 99 80 61 42 23} {
326 recorder Node -file output/node$theNode.out  -time  -node
    $theNode-dof 1 2 disp
327 }
328 recorder Element -ele 23 -time -file output/stress23.out -time
    material 2 stress
329 recorder Element -ele 41 -time -file output/stress41.out -time
    material 2 stress
330 recorder Element -ele 59 -time -file output/stress59.out -time
    material 2 stress
331 recorder Element -ele 77 -time -file output/stress77.out -time
    material 2 stress
332 recorder Element -ele 37 -time -file output/stress37.out -time
    material 2 stress
333 recorder Element -ele 37 -time -file output/strain37.out -time
    material 2 strain
334 constraints Transformation
```

```
335 numberer RCM
336 test NormDispIncr 1.E-6 25 2
337 integrator LoadControl 1 1 1 1
338 algorithm Newton
339 system BandGeneral
340 analysis Static
341 analyze 3
342 puts "soil gravity nonlinear analysis completed ..."
```

Lines 334–342 perform the gravity analysis.

Note that the command "loadConst –time 0" is not used, so the dynamic analysis begins at 3 s. Correspondingly, in the earthquake acceleration file "elcentro.txt", the first 3 s of data (the first 300 zeros in this example) are unused.

```
343 wipeAnalysis
344 constraints Transformation
345 test NormDispIncr 1.E-6 25 2
346 algorithm Newton
347 numberer RCM
348 system BandGeneral
349 integrator Newmark 0.55 0.275625
350 analysis Transient
351 set startT [clock seconds]
352 pattern UniformExcitation 1 1 -accel "Series -factor 3 -filePath
    elcentro.txt -dt 0.01"
353 analyze 2400 0.005
354 set endT [clock seconds]
355 puts " Completed time: [expr $endT-$startT] seconds."
```

Analysis Results: The nonlinear moment–curvature responses of the 3rd Gauss point in the 7th beam element and the stress–strain responses of the 37th soil element (see Fig. 6.1) are shown in Figs. 6.2 and 6.3, respectively. The model exhibits strong nonlinearity.

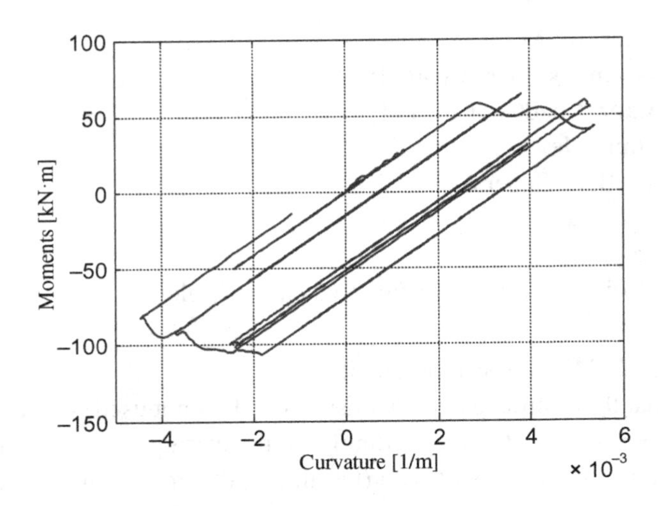

Fig. 6.2 The moment–curvature responses of the 3rd Gauss point in the 7th beam element.

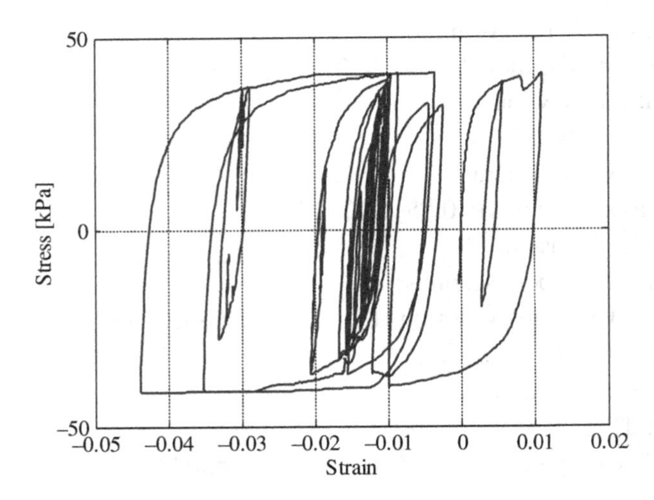

Fig. 6.3 The stress–strain responses of the Gauss point in the 37th soil element.

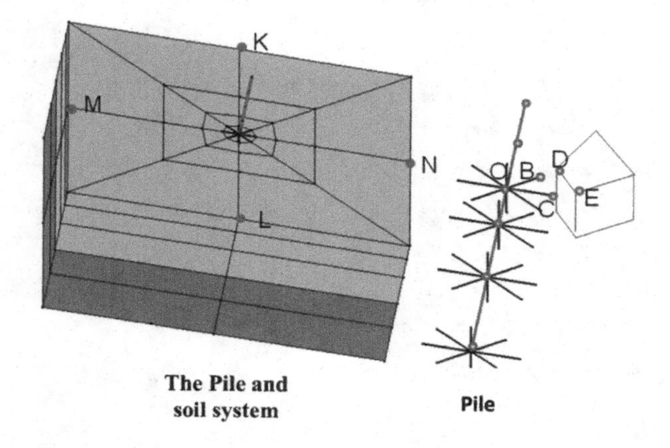

Fig. 6.4 The pile and soil system.

Note that in this example the volume of the piles is ignored. It could otherwise be simulated with a set of radial, rigid beam–column links whose length is the same as the pile radius normal to the piles' longitudinal axis. Each soil node at the periphery of a pile and the corresponding outer node of the rigid links would form a node pair to be tied together (e.g., nodes B and D in Fig. 6.4 are enforced to have the same translation displacements using "equalDOF" command). Professor Elgamal of the University of San Diego has published papers on this technique.

Reference

[1] Barbato, M., Gu, Q., Conte, J. P. Probabilistic pushover analysis of structural and geotechnical systems. *J. Struct. Eng.* (ASCE), 2010, *136*(11): 1330–1341.

Chapter 7

Fluid–Solid Coupling

Example: Dynamic Analyses of a Fluid–Solid Coupling System

(A) Brief description of the example

The model introduced in this chapter can be extended to simulate complicated dam-reservoir-foundation systems [1]. For illustrative purposes, a simple fluid–solid coupling system is shown in Fig. 7.1. The system consists of solid elements for modeling the dam, fluid elements for the water, fluid–solid interface elements, viscous boundary elements for the fluid boundary, and viscous-spring boundary elements for the foundation boundary. Though much simplified, a similar set of elements would be applied in modeling an actual dam.

The model has 16 nodes and 8 elements. The solid element models the concrete dam with nodes #1 to #8. The fluid element consists of nodes #9 to #16. The zero-thickness fluid–solid interface element consists of nodes #2, #3, #6, #7, #9, #12, #13, and #16. There are four zero-thickness viscous-spring boundary elements. One consists of nodes #3, #4, #8, and #7, the second consists of nodes #1, #2, #6, and #5, the third has nodes #1, #2, #3, and #4, and the fourth has #1, #5, #8, and #4. The viscous-spring boundary elements are along the bottom and sides of the concrete element. The fluid viscous boundary element consists of nodes #10, #11, #15, and #14, which are on the right side of the water element. Each of the bottom foundation nodes (i.e., nodes #1 through #4) is subjected to two loads along the x and z directions, whose magnitudes are both set to be large enough such that the concrete yields significantly. The truncated Drucker–Prager (DP) model is used to simulate the mechanical behaviors of concrete as shown

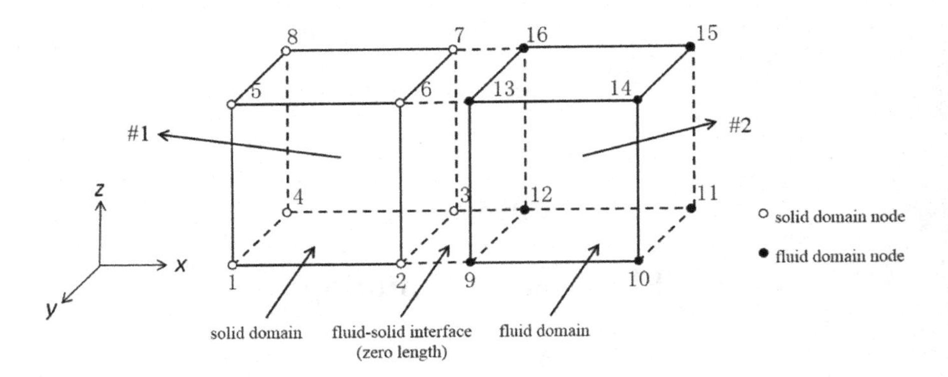

Fig. 7.1 A 3D fluid–solid coupled system.

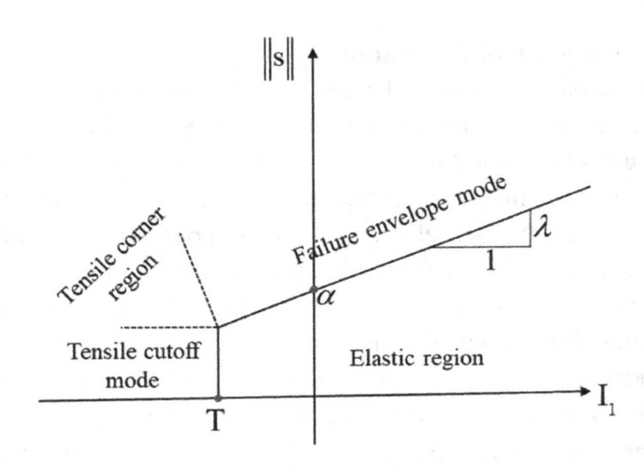

Fig. 7.2 The truncated Drucker Prager (DP) model.

in Fig. 7.2. Table 7.1 shows the properties of the solid concrete elements. The mass density of the concrete $\rho = 2400\,\text{kg/m}^3$.

The water's density is $1000\,\text{kg/m}^3$. The properties of the viscous-spring boundary element are E (its elastic modulus) $= 25\,\text{GPa}$, G (shear modulus) $= 10\,\text{GPa}$, with a rock density of $2600\,\text{kg/m}^3$. The parameter describing the normal viscous-spring boundary $\alpha_N = 1.33$ and that describing the tangential boundary $\alpha_T = 0.67$.

Table 7.1 Concrete properties assumed.

Parameter	Value	Physical meaning
G	14957 MPa	Shear modulus
K	17677 MPa	Bulk modulus
λ	0.11	The slope of the failure envelope
A	26.614 MPa	A strength-related material parameter
T	−2.0684 MPa	The hydrostatic tensile strength

(B) The Tcl commands (SI units) are as follows:

```
1  model    basic  -ndm 3  -ndf 3
2  node     1    1.00    0.00    0.00
3  node     2    1.00    1.00    0.00
4  node     3    0.00    1.00    0.00
5  node     4    0.00    0.00    0.00
6  node     5    1.00    0.00    1.00
7  node     6    1.00    1.00    1.00
8  node     7    0.00    1.00    1.00
9  node     8    0.00    0.00    1.00
10 nDMaterial  TruncatedDP  1  3  2400000  1.4957e10  1.7677e10
   0.11  2.6614e7  -2.0684e6
```

Line 10 defines a nonlinear truncated Drucker–Prager model. The numbers following "TruncatedDP" are tag of the material, dimension, density, and other parameters whose physical meanings and values are listed in Table 7.1.

```
11  set  g  9.8
12  set rhog  [expr -2400000*$g]
```

Lines 11 and 12 define the body force density for gravity loading.

```
13  element bbarBrick  1  1  2  3  4  5  6  7  8  1  0  0 $rhog
```

Line 13 defines a block element. The numbers following "bbarBrick" are tag of the element, its eight nodes, the material tag, and the body force density in the *x*, *y*, and *z* directions.

```
14  recorder Element -ele 1 -time -file stress1.out -time  material  1
    stress
```

```
15  recorder Element -ele 1 -time -file stress2.out -time  material 2
    stress
16  recorder Element -ele 1 -time -file stress3.out -time  material 3
    stress
17  recorder Element -ele 1 -time -file stress4.out -time  material 4
    stress
18  recorder Element -ele 1 -time -file stress5.out -time  material 5
    stress
19  recorder Element -ele 1 -time -file stress6.out -time  material 6
    stress
20  recorder Element -ele 1 -time -file stress7.out -time  material 7
    stress
21  recorder Element -ele 1 -time -file stress8.out -time  material 8
    stress
22  recorder Element -ele 1 -time -file strain1.out -time  material 1
    strain
23  recorder Element -ele 1 -time -file strain2.out -time  material 2
    strain
24  recorder Element -ele 1 -time -file strain3.out -time  material 3
    strain
25  recorder Element -ele 1 -time -file strain4.out -time  material 4
    strain
26  recorder Element -ele 1 -time -file strain5.out -time  material 5
    strain
27  recorder Element -ele 1 -time -file strain6.out -time  material 6
    strain
28  recorder Element -ele 1 -time -file strain7.out -time  material 7
    strain
29  recorder Element -ele 1 -time -file strain8.out -time  material 8
    strain
30  set E 2.5E10
31  set G 10E10
32  set rho 2600
33  set R 475
34  set alphaN 1.33
35  set alphaT 0.67
```

Lines 30–35 define the material parameters: the elastic modulus, the shear modulus, the mass density, distance of scattering wave to boundary, and parameters of the viscous-spring boundary in the normal and tangential directions.

```
36  element VS3D4 5 3 4 8 7 $E $G $rho $R $alphaN $alphaT
37  element VS3D4 6 1 2 6 5 $E $G $rho $R $alphaN $alphaT
38  element VS3D4 7 1 2 3 4 $E $G $rho $R $alphaN $alphaT
39  element VS3D4 8 1 5 8 4 $E $G $rho $R $alphaN $alphaT
```

Lines 36–39 define the four-node 3D viscous-spring boundary elements used to simulate the truncated boundary on the sides and bottom of the concrete.

```
40  recorder Node -file node_5.out -time -precision 16 -node 5 -dof
    1 disp
41  recorder Node -file node_6.out -time -precision 16 -node 6 -dof
    1 disp
```

Turning now to the water,

```
42  model basic -ndm 3 -ndf 1
43  node    9   1.00  1.00  0.00
44  node   10   1.00  2.00  0.00
45  node   11   0.00  2.00  0.00
46  node   12   0.00  1.00  0.00
47  node   13   1.00  1.00  1.00
48  node   14   1.00  2.00  1.00
49  node   15   0.00  2.00  1.00
50  node   16   0.00  1.00  1.00
51  set water 2
52  nDMaterial AcousticMedium $water 2.07e9 1000.0
```

Lines 51 and 52 define an acoustic medium. The numbers following "AcousticMedium" represent tag of the material, its volume modulus, and its mass density.

```
53  element AC3D8 3 9 10 11 12 13 14 15 16 $water
54  element ASI3D8 2 2 3 6   7  9 12 13 16
55  element AV3D4 4 10 11 15 14 $water
```

Lines 53–55 define the fluid element, the fluid–solid interface element, and the fluid viscous boundary element, respectively.

```
55    fix    13    1
56    fix    14    1
57    fix    15    1
58    fix    16    1
```

Lines 55–58 constrain the water pressure of nodes 13–16 to be zero.

```
59  recorder Node -file node_9.out -time -precision 16 -node 9 -dof
    1 2 3 disp
60  recorder Node -file node_10.out -time -precision 16 -node 10 -dof
    1 disp
61  puts ''Define model acoustic ok....''
62  timeSeries Path 1 -dt 0.01 -filePath elcentro.txt -factor 1.0
```

Line 62 defines time series with tag 1. The data are the acceleration vector read from the file "elcentro.txt" multiplied by the factor 3.0, and the time interval between two adjacent accelerations is 0.01 s.

```
63  pattern Plain 111 1 {
64    load        1 5e6  0 5e6
65    load        2 5e6  0 5e6
66    load        3 5e6  0 5e6
67    load        4 5e6  0 5e6
68  }
```

Lines 63–68 define a loading pattern with tag 111. The external forces have a magnitude of 5×10^6 N multiplied by time series of tag #1, and they are applied at nodes #1 through #4 along both the x and z directions. For the actual dam foundation system, the external forces are usually applied on the boundaries of the foundations.

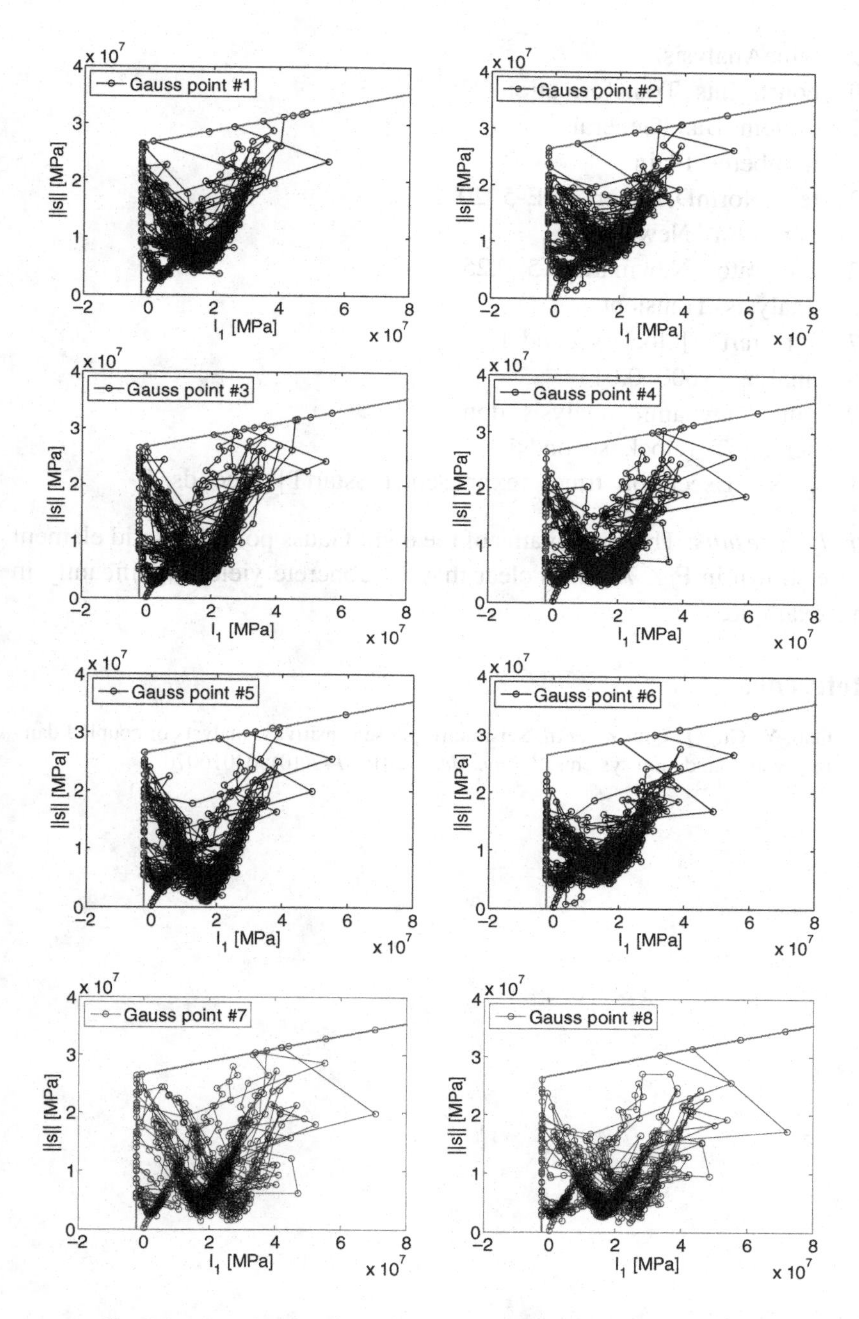

Fig. 7.3 Stress paths of the eight Gauss points of the solid element.

```
69 wipeAnalysis
70 constraints Transformation
71 system BandGeneral
72 numberer Plain
73 test NormDispIncr 1.0E-5 20 2
74 algorithm Newton
75 integrator Newmark 0.5 0.25
76 analysis Transient
77 set startT [clock seconds]
78 analyze 1600 0.01
79 puts ''Dynamic analysis done...''
80 set endT [clock seconds]
81 puts ''Execution time: [expr $endT-$startT] seconds.''
```

Analysis results: The stress paths of the eight Gauss points of solid element 1 are shown in Fig. 7.3. It is clear that the concrete yields significantly in this example.

Reference

[1] Gao, Y., Gu, Q., Qiu, Z., *et al.* Seismic response sensitivity analysis of coupled dam-reservoir-foundation systems. *J. Eng. Mech.* 2016, *142*(10): 04016070.

Chapter 8

Numerical Analyses of Soil Liquefaction

8.1 Example 1: Soil Liquefaction Analyses Using a Bounding Surface Model

(A) Brief description

When saturated sandy soils are subjected to earthquakes, the pore water pressure may increase and the soil's effective stress may decrease. The soil particles lose contact and inter-particle friction drops. The soil may lose part of its strength and stiffness and even behave like a liquid, a process which is called soil liquefaction. Soil liquefaction may be extremely harmful, causing foundations to settle down or even the buildings and bridges to collapse. To simulate soil liquefaction is of significant importance. For simplicity's sake, a four-node quadrilateral (or quad) element with liquefiable soil material is taken as an example, as shown in Fig. 8.1.

The soil is fully saturated and undrained, with the water level at the ground surface, i.e., at the same depth as nodes #3 and #4 in Fig. 8.1. The force F shown in Fig. 8.2 is applied to nodes 3 and 4.

More detailed descriptions of the model can be found in [1] and [2]:

(B) Tcl commands used are as follows:

```
1    model basic -ndm 2 -ndf 3
```

In line 1, "-ndf 3" specifies that each node has three dof (i.e., two translational dofs and the third dof denoting an integral over time of the pore water pressure).

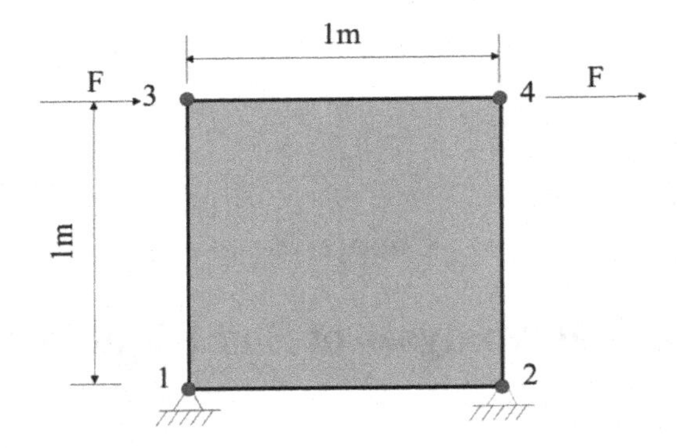

Fig. 8.1 The FE model of liquefiable soil.

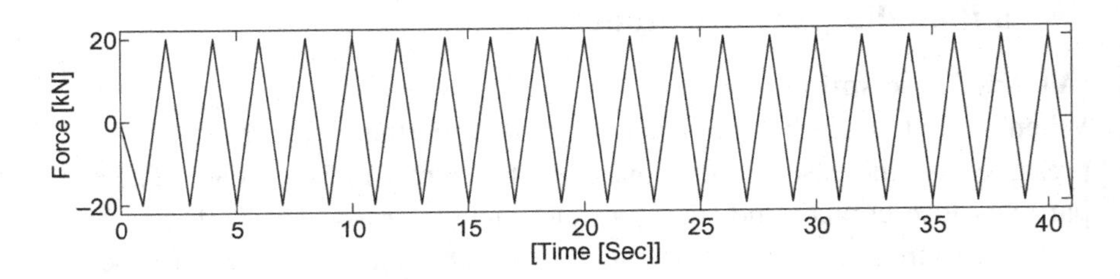

Fig. 8.2 Cyclic loading history.

```
2   node 1  0.0 0.0
3   node 2  1.0 0.0
4   node 3  0.0 1.0
5   node 4  1.0 1.0
6   nDMaterial BoundingSurfaceSand  1  2  1.90  0.5066  2000.0
    0.818  0.05  0.5423  0.75  0.0  1.0  1.9  0.1811  101.3250
    1.01537  0.934  0.019  0.7  3.5  0.75  0.4  0.0  0.0
```

Line 6 defines the bounding soil's surface plasticity. The numbers following "BoundingSurfaceSand" are the material's tag, the number of dimensions for the plane–strain analysis, the saturated soil's mass density (assumed to be 1.90 t/m^3), and the cutoff of confinement close to liquefaction (assumed to be 0.5066 Pa). The shear modulus is at a reference pressure of 2000 kPa. An initial void ratio of 0.818 will be assumed with a Poisson's ratio of 0.05 for the plasticity analysis. The parameters 0.5423, 0.75, 0.0, 1.0, and 1.9 are model constants defined in variables accounting for monotonic and cyclic loading. They affect the changes in the plastic bulk modulus. Among them, 0.75 denotes the ratio of the stress ratio at the phase transformation surface to the stress ratio at the failure surface. The parameter 0.1811 represents the change in the plastic shear modulus. 101.3250 kPa is the standard atmospheric pressure. The stress ratio at failure surface based on triaxial tests is 1.01537. The parameters 0.934, 0.019, and 0.7 define the critical-state line. 3.5 and 0.75 are model constants used to calibrate the phase transformation line. An elastic Poisson's ratio of 0.4 under elastic gravity analysis (considering anisotropy) was assumed. The last two parameters signify that the initial confinement is 0 and the initial $k_o = 0$.

```
7   element  quadUP  1  1  2  4  3  1.0  1  2.2e6  1  5.09e-8
    5.09e-8  0.0  -480.0  0
```

Line 7 defines a solid–fluid fully-coupled plane-strain element with tag 1 and four element nodes with tags of 1, 2, 4, and 3 in counter-clockwise order around the element. The thickness of the element is 1.0 m. The material tag is 1 (previously defined). The bulk modulus of the fluid phase is 2.2×10^6 kPa for water. 5.09×10^{-8} m/s is the permeability in both the horizontal and vertical directions. The optional gravity acceleration components in horizontal and vertical directions are 0.0 and $-480\,\text{m/s}^2$, assumed so in order to achieve a desirable initial confinement level in this example. (The default is $0\,\text{m/s}^2$.) The last number 0 is an optional uniform element normal traction.

```
8   recorder Element -ele 1 -time -file stress.out material 1 stress
9   recorder Element -ele 1 -time -file strain.out material 1 strain
10  fix 1 1 1 0
11  fix 3 0 0 1
12  fix 2 1 1 0
13  fix 4 0 0 1
```

Lines 10–13 define the constraints. The horizontal and vertical degrees of freedom of nodes 1 and 2 are fixed, while the pressure (the 3rd dof) of nodes 3 and 4 is set to zero.

```
14  equalDOF 3 4 1 2
```

Line 14 constrains the displacements of nodes 3 and 4 to be identical along both the *X* and *Y* directions to simulate simple shear.

```
15  numberer RCM
16  system ProfileSPD
17  test NormDispIncr 1.0e-8 50 1
18  algorithm KrylovNewton
19  constraints Transformation
20  integrator Newmark 1.5 1.
21  analysis VariableTransient
22  analyze 5 5.0e3 [expr 5.0e3/100] 5.0e3 20;
```

Lines 20–22 perform a "static" gravity analysis by using transient (dynamic) analysis. The time step is set to be large enough such that the system quickly reaches a stable equilibrium, and special parameters are used in Newmark's time-step integration.

```
23  updateMaterialStage -material 1 -stage 1000
24  analyze 3 5.0e3 [expr 5.0e3/100] 5.0e3 20;
```

Lines 23 and 24 change the state of the material from linear elastic to plastic and continue to perform the "static" gravity analysis to reach a new stable equilibrium.

Note: Lines 10–14 are common boundary conditions for soil liquefaction problems and can be extended to large-scale computational models. Note that the water's degree of freedom in the UP element is not water pressure, but the integral of the pressure over time (in order to obtain a symmetric stiffness matrix). Details can be found in O.C. Zienkiewicz's book on computational geomechanics. Fixing the 3rd dof enforces the integral of the water pressure to be zero. Specifying other forms of pressure boundary is not so easy, although this is possible (e.g., a multi-support loading pattern). Lines 20–22 perform dynamic analysis for the gravity loading and the material is linear at that moment. After that the material is changed to nonlinear using the command "updateMaterialStage" in line 23 and it is re-analyzed in line 24. Dynamic analysis can then be performed.

The material model “BoundingSurfaceSand” may not be found in the official website of OpenSees, but may be found at https://github.com/OpenSeesXMU. Similarly, in the later chapters, some unavailable models may be found in https://github.com/OpenSeesXMU.

```
25  wipeAnalysis
```

Line 25 deletes the above analysis and begins to perform the cyclic loading analysis.

```
26  set P_max 20.0
27  pattern Plain 1 ‘‘Series -time {40000.0 40001.0 40002.0 40003.0
    40004.0 40005.0 40006.0 40007.0 40008.0 40009.0 40010.0
    40011.0 40012.0 40013.0 40014.0 40015.0 40016.0 40017.0
    40018.0 40019.0 40020.0 40021.0 40022.0 40023.0 40024.0
    40025.0 40026.0 40027.0  0028.0 40029.0 40030.0 40031.0
    40032.0 40033.0 40034.0 40035.0 40036.0 40037.0 40038.0
    40039.0 40040.0 40041.0 } -values { 0.0 -1.0 1.0 -1.0 1.0 -1.0
    1.0 -1.0 1.0 -1.0 1.0 -1.0 1.0 -1.0 1.0 -1.0 1.0 -1.0 1.0 -1.0 1.0
    -1.0 1.0 -1.0 1.0 -1.0 1.0 -1.0 1.0 -1.0 1.0 -1.0 1.0 -1.0 1.0 -1.0
    1.0 -1.0 1.0 -1.0 1.0 -1.0 }’’ {
    load 3  $P_max  0.0 0
    load 4  $P_max  0.0 0
    }
```

Lines 26 and 27 define the loading pattern shown in Fig. 8.2. A force of 20 kN is applied to both nodes 3 and 4 in the x direction.

```
28  constraints Transformation
29  test NormDispIncr 1.0e-8 50 0
30  numberer RCM
31  algorithm Newton
32  system BandGeneral
33  rayleigh 0.0 0.0 0.02 0.0
34  integrator Newmark 0.6 [expr pow(0.6+0.5, 2)/4]
35  analysis Transient
36  analyze 4000 0.01
```

Lines 28–36 perform the dynamic analysis assuming the cyclic loading of Fig. 8.2.

Analysis results: Figs. 8.3 and 8.4 show part of the responses of Gauss point #1.

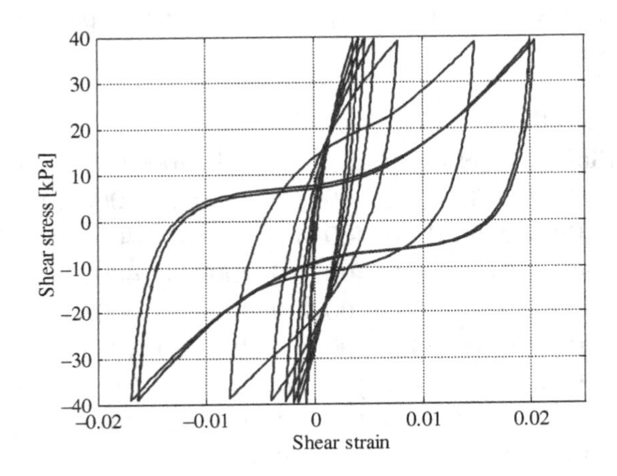

Fig. 8.3 Shear stress–shear strain responses of Gauss point 1.

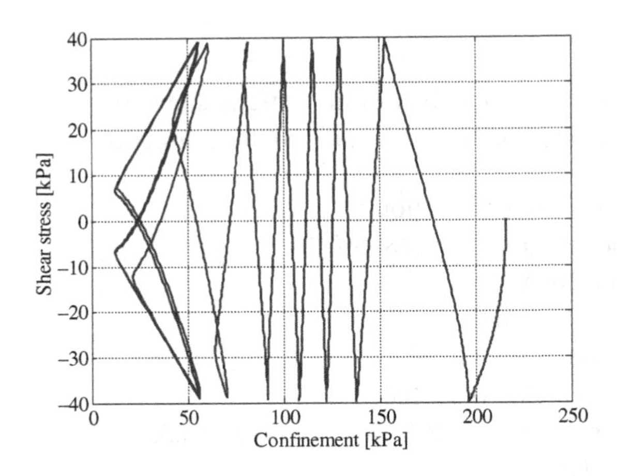

Fig. 8.4 Shear stress at Gauss point 1 as a function of confinement.

8.2 Example 2: Soil Liquefaction Analysis Using a Multi-yield Surface J2 Plasticity Model

(A) Description of the problem is as follows:

The model is the same as that in Example 1, except that the material is changed from "BoundingSurfaceSand" to "PressureDependMultiYield".

(B) Tcl commands include the following:

```
1   nDMaterial PressureDependMultiYield 1 2 2 6.e4 2.4e5 31 0.1
    80 0.5 26.5 0.1 0.2 5 10 0.015 1.
```

Line 1 defines a plastic soil material. The numbers following "PressureDependMultiYield" are the material tag, the number of dimensions, the saturated soil's mass density, the low-strain shear modulus, the bulk modulus, the friction angle at peak shear strength, the octahedral peak shear strain, and the reference mean effective confinements. "0.5" is the parameter defining the relationship between the shear (and volume) modulus and the confinements, "26.5" denotes a phase transformation angle in degrees. "0.1" is the constant defining the soil's contraction rate or pore pressure buildup. "0.2" and "5" are parameters related to the dilation.

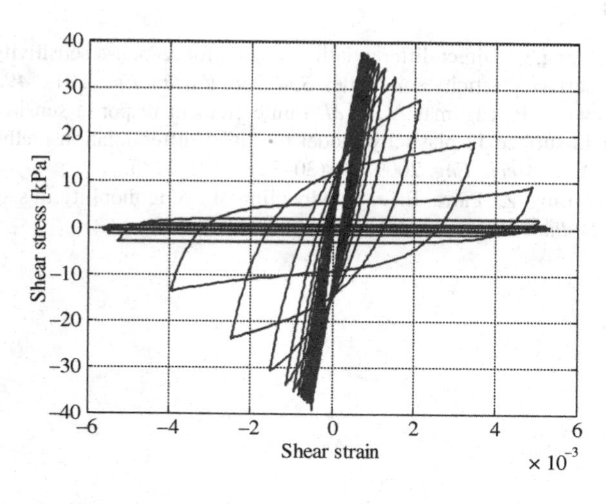

Fig. 8.5 Shear stress–shear strain responses of Gauss point 1.

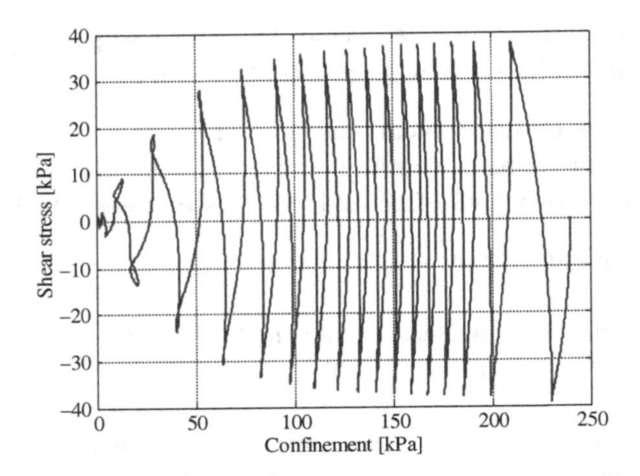

Fig. 8.6 The shear stress of Gauss point 1 at various confinements.

"10", "0.015", and "1" are liquefaction-related parameters. A more detailed description of the material can be found in [3].

Analysis results: Figs. 8.5 and 8.6 show part of the responses of Gauss point 1.

References

[1] Gu, Q. and Wang, G. Direct differentiation method for response sensitivity analysis of a bounding surface plasticity soil model. *Soil Dyn. Earthq. Eng.* 2013, *49*: 135–145.

[2] Gu, Q., Conte, J. P., Elgamal A., *et al.* Finite element response sensitivity analysis of multi-yield-surface J2 plasticity model by direct differentiation method. *Comput. Methods in Appl. Mech. Eng.* 2009, *198*(30–32): 2272–2285.

[3] Elgamal, A., Yang, Z., Parra, E., *et al.* Modeling of cyclic mobility in saturated cohesionless soils. *Int. J. Plast.* 2003, *19*(6): 883–905.

Chapter 9

Numerical Optimization

Numerical optimization is commonly used in various subfields of civil engineering such as structure reliability analysis, FE model updating, structure identification, and structure optimization. This chapter introduces a nonlinear optimization software package (SNOPT) developed by Professor Philip Gill and co-workers, which has been integrated into OpenSees. The OpenSees–SNOPT framework is general and flexible and can be used to solve a wide range of FE-based optimization problems in structural and geotechnical engineering [1].

9.1 SNOPT Optimization

The basic process for SNOPT optimization is as follows.

As Fig. 9.1 shows, SNOPT analysis consists of six steps. Step 1 creates an FE model which keeps the Tcl in memory during the optimization. That step is optional and is necessary for FE model-based optimization. Steps 2 and 3 define the design variable (DV) and a DV-positioner using the Tcl commands “designVariable” and “designVariablePositioner”. The DV-positioner maps the physical parameters of the model (for example, an element’s elastic modulus) to the DVs. Step 4 defines a Tcl global variable to store the value of the objective function (OF) using the “objectiveFunction” command. Furthermore, Tcl routine for calculating the OF is specified (“F.tcl” in this example). Step 5 defines Tcl global variables to store the value of the constraint function (CF) using the “constraintFunction” command. Furthermore, a Tcl routine for calculating the CF is specified

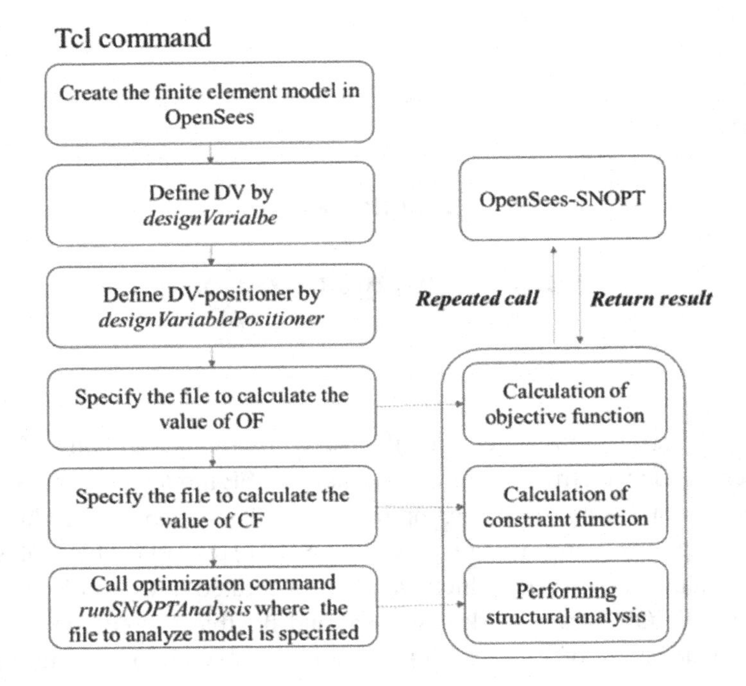

Fig. 9.1 SNOPT optimization flowchart.

(in this example, “G.tcl”). Step 6 performs the optimization by using the “runSNOPTAnalysis” command and a Tcl routine for the FE analysis is specified. OpenSees calls these functions repeatedly to run analyses and calculate the OF and CF as needed.

It is worth mentioning that Steps 1 and 3 may be omitted for general purpose optimization (rather than FE-based optimization), which is relatively easier for users (although less efficient) as shown in the example in the later part of this chapter.

The commands for optimization analysis are explained as follows.

• Command “designVariable”

designVariable $gradNumber -name $nameDV -startPt $startPoint -lowerBound $lBound -upperBound $upperBound

$gradNumber: the number of the design variable;

$nameDV: The unique Tcl variable name given to that DV by the user;

$startPoint: The initial point at the start of the optimization (i.e., the initial value of the optimization variable);

$lBound: A lower bound on the DV;

$upperBound: An upper bound on the DV.

• Command "designVariablePositioner"

designVariablePositioner $pos -dvNum $gradNumber -element 1 -sectionAggregator 2 -section 1 -material 1 $h

"designVariablePositioner" specifies the model parameters corresponding to the design variables. Such as:

$pos: Identifies the design parameter, which must start from 1 and be continuously numbered by adding 1 for each parameter; the same design variable can correspond to multiple parameters. $pos is numbered over the entire model;

$gradNumber: The number of the design variable, which is created in advance by the command "***designVariable***";

$ele: The number of the element, which is created in advance by the command "***element***" in the finite element model;

$secAgg: In the finite element model, the number of sections created in advance by the command "***section Aggregation***";

$mat: In the finite element model, the number of the material created in advance by the command "***material***";

$h: The design variable defined in Tcl. It is a string variable.

• Command "objectiveFunction"

objectiveFunction $tag -name $nameF -tclFile $tclFile -lowerBound $lBound -upperBound $uBound < -gradientName $gradF >

$tag: The tag of the objective function, which starts from 1. The current SNOPT version can only have one objective function;

$nameF: The name of the objective function, which is a unique global variable in Tcl;

$tclFile: The Tcl file provided by the user, which is used to calculate the objective function and save it as the global variable *$nameF*. With the

command ***"-gradientName $gradF"*** the file can also be used to calculate the gradient of the objective function;

$lBound: A lower bound on the objective function;

$upperBound: An upper bound for the objective function;

$gradF: The global variable name that stores the gradient of the goal function. It can be an array, but it is optional. If this function is selected, the global gradient variable must also be given in the *$tclFile* calculation.

• Command "constraintFunction"
constraintFunction $tag -name $nameG < -gradientName $gradG > -tclFile $tclFile -lowerBound $lBound -upperBound $uBound

$tag: The tag of the constraint function beginning from 1. SNOPT can have multiple constraints;

$nameG: The unique name of the constraint function;

$gradG: The global variable name (which can be a matrix) that stores the constraint function's gradient. It is optional. If this function is selected, the global constraint gradient variable must also be given in the *$tclFile* calculation;

$tclFile: A Tcl file provided by the user used to calculate the constraint function and store its value in the global variable *$nameG*. The ***"-gradientName $gradG"*** can be used to calculate the gradient of the constraint function and to save the value as the global variable ***$gradG***;

$lBound: a lower bound on the constraint function;

$upperBound: An upper bound on the constraint function;

If the number of constraints is 2, the expression can be written as follows:

array set uBound {1 4.0 2 5.0}
array set lBound {1 -1e20 2 -1e20}

That specifies that the upper and lower bounds on constraint function 1 are 4.0 and -10^{20}. The upper and lower bound on constraint function 2 are 5.0 and -10^{20}.

• Command "runSNOPTAnalysis"

runSNOPTAnalysis -maxNumIter $maxNumber -printOptPointX $resultFile -printFlag $printFlag < -tclFileToRun $tclFile >

$maxNumber: The maximum number of iterations during the optimization process;

$resultFile: The file that saves the optimization results;

$printFlag: Output signs. Different values have different meanings. A zero means the results of the analysis will not be displayed on the screen or stored in the file. When the value is 1, the analysis results of each step will be displayed but not stored. A 2 means the result of each step will be displayed on the screen and the important information needed for the start of another analysis will be stored in the file.

$tclFile: The Tcl file containing the analysis information in the FE analysis.

• Command "updateParameter"

updateParameter –dv $numDV –value $newValue

$numDV: The number of the DV whose value needs to be updated;

$newValue: The new value assigned to that DV.

9.2 An Example of the Optimization Analysis

(A) Brief description

This section uses a simple beam example to illustrate how to use SNOPT optimization. The setup is shown in Fig. 9.2.

The beam's elastic moduli (E1, E2), stiffness ratios (B1, B2), and yield strengths (F1, F2) are the model's DVs. The horizontal displacement of node 4 is the response to be calibrated. The OF is defined as the difference between the FE-predicted and the experimentally determined responses of node 4. That is,

$$F = \frac{1}{2}\sum_{i=1}^{\text{nSteps}}\left(u_i^{FEM} - u_i^{Exp}\right)^2$$

Fig. 9.2 The simple beam FE model.

Table 9.1 The initial values and the bounds of the DVs.

	***E*1 (Pa)**	***E*2 (Pa)**	***B*1 (-)**	***B*2 (-)**	***F*1 (Pa)**	***F*2 (Pa)**
Maximum value	1e20	1e20	1e20	1e20	1e20	1e20
Minimum value	1e8	1e8	0.0	0.0	1e5	1e5
Initial value	1.8e8	1.8e8	0.016	0.016	2.7e5	2.7e5

where u_i is the horizontal displacement of node 4 at the ith step, and the superscripts "*FEM*" and "*Exp*" denote the FEM's predictions and experimentally determined values. In this example the so-called experimental responses are not measured data, but responses calculated by OpenSees using another set of parameters. The upper bound, lower bound, and the initial values of the DVs are shown in Table 9.1.

The optimization requires three separate Tcl files: the main file "main.tcl", the file for structural analysis "tclFileToRun", and the file to calculate the objective function "F.tcl". Since there are no constraints in this example, the "constraint" command and its tcl file are not needed. In addition, this example does not use the command "designVariable-Positioner", instead wiping and re-building the FE model whenever there are updated DVs. In this way, the process is simplified. In addition, SNOPT comes with a SPECS file "sntoya.spc", which allows users to control the optimization by, for example, adjusting the calculation errors.

(B) Tcl Commands included are as follows:

1 optimization

The first line is to create an optimized domain.

2 designVariable 1 -name DV_E1 -startPt 1.8e8 -lowerBound 1.0E8 -upperBound 1.e20

3 designVariable 2 -name DV_fy1 -startPt 270000. -lowerBound 100000.0 -upperBound 1.e20

4 designVariable 3 -name DV_b1 -startPt 0.016 -lowerBound 0.0 -upperBound 1.e20

```
5  designVariable 4 -name DV_E2 -startPt 1.8e8 -lowerBound
   1.0E8 -upperBound 1.e20
6  designVariable 5 -name DV_fy2 -startPt 270000. -lowerBound
   100000.0 -upperBound 1.e20
7  designVariable 6 -name DV_b2 -startPt 0.016 -lowerBound
   0.0 -upperBound 1.e20
```

Lines 2–7 define the DVs. In line 2, the number of the DV is 1 and the global variable name in Tcl is "*DV_E1*". The initial value of the optimization variable is 1.8×10^8, and the lower and upper bounds are 10^8 and 10^{20} (i.e., infinite).

```
8  objectiveFunction 1 -name F -tclFile F.tcl -lowerBound -1.e20
   -upperBound 1.e20;
```

Line 8 defines the OF. The number of the OF is 1 and its unique global variable name is "*F*". The Tcl file provided by the user is F.tcl (though the name of the Tcl file is not necessarily the same as the name of the OF). The OF's lower bound is -10^{20} and the upper bound is 10^{20}.

```
9  runSNOPTAnalysis -maxNumIter 50 -printOptPointX OptX.out -
   printFlag 1 -tclFileToRun tclFileToRun.tcl
```

Line 9 is the command for running an SNOPT analysis. "-maxNumIter 50" indicates that the maximum number of iterations is 50. "-printOptPointX OptX.out" indicates that the optimization results are saved in the file OptX.out. "-printFlag" represents the output flag: if the following number is "1", the results will be reported, if the number is "0", they will not. "-tclFileToRun tclFileToRun.tcl" indicates that the running file is "tclFileToRun.tcl".

The file **F.tcl** used to calculate the value of the objective function "*F*" will be called automatically by OpenSees–SNOPT after the structure is analyzed.

```
1  set F 0
2  set fileId_exp [open ''node4_exp.txt'' ''r'']
3  set fileId_fem [open ''node4.out'' ''r+'']
```

The files "node4_exp.txt" and "node4.out" are the experimentally determined and FE-predicted responses of node 4, respectively.

```
4    while {[gets $fileId_exp u_exp_line] >= 1} {
5       if {[gets $fileId_fem u_fem_line] >= 1} {
6             set count [scan $u_exp_line ''%f %e'' time_exp u_exp ]
7             if {$count != 2} {
8                error ''Error reading input - terminating script''
9                exit;
10            }
11          set count [scan $u_fem_line ''%f %e %e'' time_fem u_fem
            u_tmp]
12            if {$count != 3} {
13               error ''Error reading input - terminating script''
14               exit;
15            }
16      set F [expr $F + ($u_exp-$u_fem)*($u_exp-$u_fem)];
17      } ; #if
18   } ; #wihle
19   close $fileId_exp
20   close $fileId_fem
```

The file **tclFileToRun.tcl** will analyze the structural response after the DVs have been updated.

```
1    wipe
```

The first line removes the FE model in OpenSees (including elements, nodes, etc.) from the Tcl memory, however, the previously defined Tcl variables (DV_E1, DV_fy1) are not removed.

```
2    model basic -ndm 2 -ndf 2
3    node 1   0.0 0.0
4    node 2  10.0 0.0 -mass 3.1741e+003  0 0.0
5    node 3  20.0 0.0 -mass 4.1741e+003  0 0.0
6    node 4  30.0 0.0 -mass 5.1741e+003  0 0.0
7    fix 1 1 1
8    fix 2 0 1
```

```
9   fix 3 0 1
10  fix 4 0 1
11  uniaxialMaterial Steel01  1  $DV_fy1 $DV_E1 $DV_b1
12  uniaxialMaterial Steel01  2  $DV_fy2 $DV_E2 $DV_b2
```

Lines 11–12 define two uniaxial materials. The three variables following Steel01 are the DVs.

```
13  element truss  1  1 2  0.01  1
14  element truss  2  2 3  0.02  2
15  element truss  3  3 4  0.02  1
16  timeSeries Path 1 -dt 0.02 -filePath tabas.txt -factor 9.8
```

Line 16 defines the load path.

```
17  pattern UniformExcitation 1 1 -accel 1
```

Line 17 defines the form of the load.

```
18  constraints Plain
19  numberer RCM
20  test NormDispIncr 1.e -4 25 0
21  algorithm Newton
22  system BandSPD
23  integrator Newmark 0.55 0.275625
24  analysis Transient
25  recorder Node -file node4.out -time -node 4 -dof 1 2 -precision
    16 disp
26  set ok [analyze 2000 0.01]
```

Line 26 is set to run 2000 steps with the time step 0.01 s. If it succeeds, it returns 0 (i.e., the value of the Tcl variable "*ok*" is 1); otherwise it returns -1.

```
27  remove recorder
```

Line 27 deletes the previously defined recorders. This is necessary to close the recorder files.

```
28  return $ok
```

The SPECS file is the default SPECS file for customizing SNOPT by specifying various options.

The **sntoya.spc** optimization control utility. A detailed description can be found in [2].

```
Begin Toy NLP problem
Major Print level      000001
*                     (JFLXBT)
Minor print level        1
Solution               yes
Major feasibility tolerance 1.0e-6 * target nonlinear constraint
violation
Major optimality tolerance 1.0e-6 * target complementarity gap
Minor feasibility tolerance 1.0e-6
System information          Yes
New superbasics            10000
Hessian          full memory
End Toy NLP problem
```

More details are provided in the SNOPT user manual.

Analysis Results: Table 9.2 shows the initial values, bounds, and optimization results of the DVs in this example.

Figure 9.3 compares the "experimental" responses with the FE predictions before and after optimization.

Table 9.2 The initial values, bounds, and optimization results of the DVs.

Design variable	Real value	Upper bound	Lower bound	Initial value	Optimization result
E1 (Pa)	2.01e8	1.e20	1.e8	1.8e8	2.28e8
fy1 (Pa)	307,460	1.e20	1.e5	2.7e5	3.09e5
b1	0.02	1.e20	0.0	0.016	1.51e-2
E2 (Pa)	1.05e8	1.e20	1.e8	1.8e8	1.20e8
fy2 (Pa)	206,460	1.e20	1.e5	2.7e5	2.196e5
b2	0.04	1.e20	0.0	0.016	1.65e-2

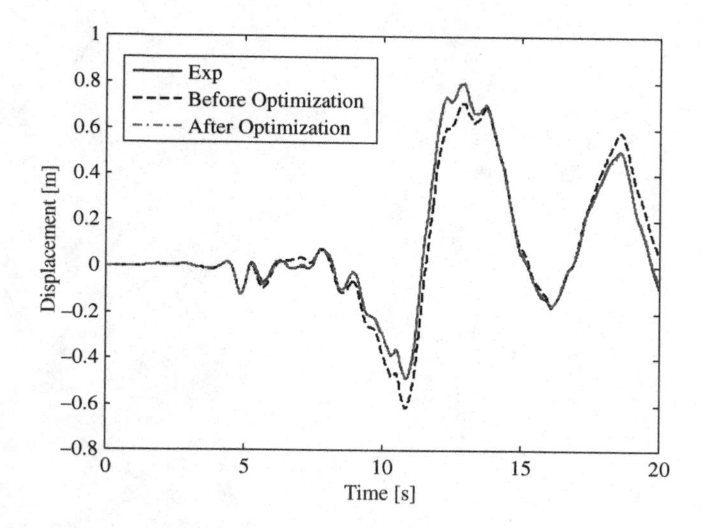

Fig. 9.3 Comparison of the experimental and FE-predicted response.

References

[1] Gu, Q., Barbato, M., Conte, J. P., *et al.* OpenSees-SNOPT framework for finite-element-based optimization of structural and geotechnical systems. *J. Struct. Eng.* 2012, *138*(6): 822–834.

[2] Philip, E. G., Murray, W., and Michael, A. S. User's Guide for SNOPT Version 7: Software for Large-Scale Nonlinear Programming. 2008.

Chapter 10

Coupling OpenSees with Other Software Using the Client–Server Technique

> *Note*: In many practical engineering problems, OpenSees is a "structural simulation tool" like a calculator, which will be called in by higher-level algorithms when needed. In such cases, the client–server (CS) integration technique can be very useful.

Let us take as an example the OpenSees–SNOPT optimization. At each iteration, the "upper-level" optimization algorithm determines the design variables (DVs), calls on OpenSees to perform a specified analysis, extracts the analysis results (such as the structure's responses), and calculates the objective and constraint functions. SNOPT then gets those values, calculates the new DVs, and calls OpenSees again for the next iteration. That process is repeated until the optimum is attained.

In that process OpenSees needs to remain in memory and await commands from SNOPT. It performs each command and sends back the analysis results to SNOPT. After this, OpenSees continues to wait for new commands. OpenSees behaves like a "calculator server", and SNOPT uses this calculator by handling and manipulating a simple "client" amounting to a few lines of Tcl commands (which may be integrated into other software platforms). The one client to one server (CS) technique is developed based on this idea. More details may be found in [1].

OpenSees.exe can be run to create a server by sourcing a tcl file server.tcl. Then, another OpenSees.exe can be run to source the client.tcl to

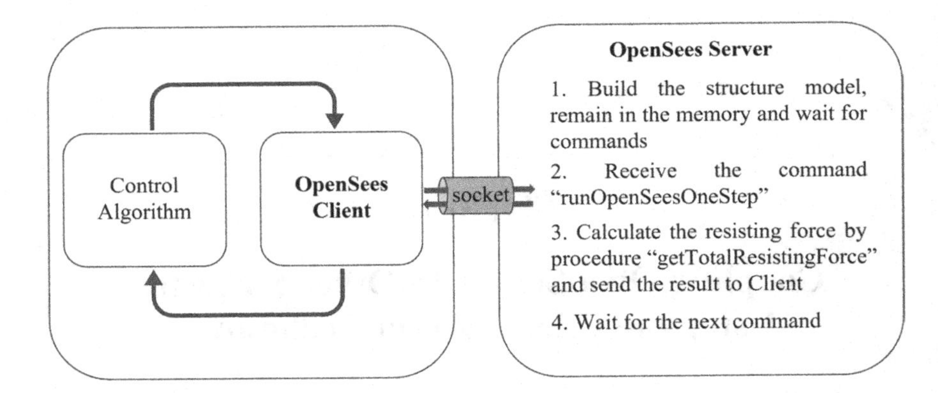

Fig. 10.1 A schematic diagram of CS technique.

build the client terminal. The server and client are connected through a network "socket". The two OpenSees.exe files run simultaneously. The client can send commands to the server and the server sends back the results to the client through the socket. The two computers (or maybe one computer having two programs running) both require the TCP/IP or other internet protocols.

The example consists of 3 files: server.tcl, client.tcl, and model.tcl. A schematic diagram of the CS technique is shown in Fig. 10.1:

(1) Server side

The following Tcl codes are used to create a server which will build the structure model and await commands sent from the client terminal (Fig. 10.1). Whenever the server receives commands from the client through the socket, it will run them and send the results back to the client.

```
S1  proc accept {sock ip port} {
S2    fconfigure $sock -blocking 1 -buffering none
S3    fileevent $sock readable [list respond $sock]
S4  }
S5  proc respond {sock} {
S6      if {[eof $sock] || [catch {gets $sock data}]} {
S7          close $sock
S8          }
```

```
S9      else {
S10       # eval $data;
S11       #global Fx;
S12       #global My;
S13       # getTotalResistingForce ;
S14       #puts $sock " $Fx, $My"
S15       puts $sock $data
S16       return
S17     }
S18 }
S19 # source model.tcl
S20 socket -server accept 7200
S21 vwait forever
```

Lines S1–S4 define a procedure "accept" for receiving commands from the client. The commands in $sock are sent to another procedure "respond" to be dealt with as defined in Lines S5–S18. Line S20 creates a server socket and S21 defines an event loop to wait for the commands from the client.

(2) Client

A client can be built by running another OpenSees and exploiting the following Tcl codes. Generally, these client codes are integrated into and called by other software platforms (e.g., Matlab Simulink). Lines C1–C5 establish a socket connection to the server. "localhost" may be used if the server is in the same computer, or the IP of the server may be used. The number after "localhost" is the port.

```
C1  set s [socket localhost 7200] ;
C2  fconfigure $s -buffering none ;
C3  puts $s "runOpenSeesOneStep 1 0.01 0.005 0.0002 "
C4  set a [gets $s]
C5  puts $a
```

Line C3 sends the command string "runOpenSeesOneStep 1 0.01 0.005 0.0002" to the server through the socket "$s". Line C4 receives the results, which in this example is the same string that was sent to the server (i.e., "runOpenSeesOneStep 1 0.01 0.005 0.0002", see S15).

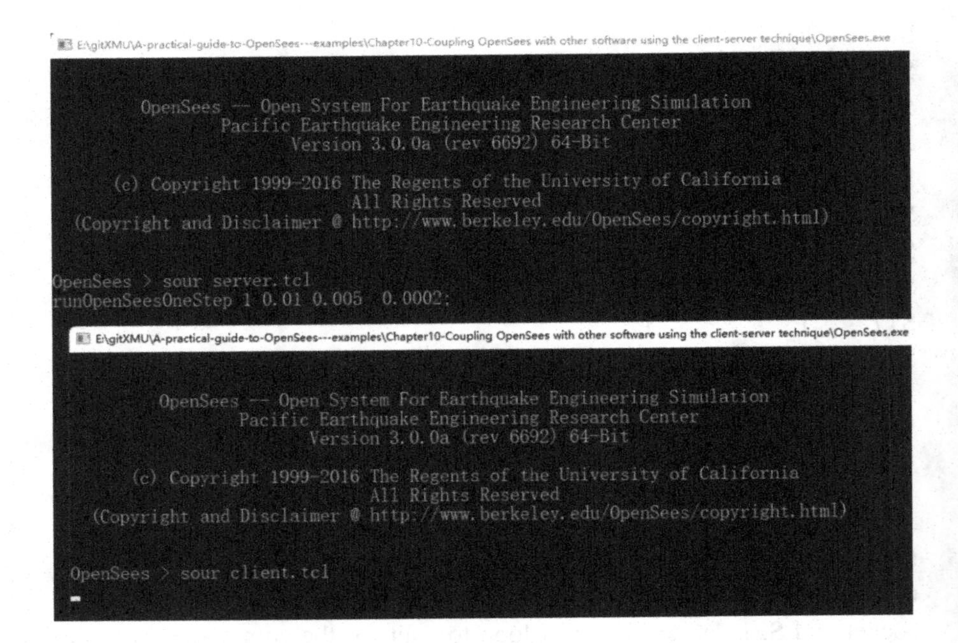

Fig. 10.2 An example of the use of CS technique.

The result is shown in Fig. 10.2.

Note: Establishing a connection between the client and the server socket is relatively easy as long as the two computers are connected through a network which requires the TCP/IP or other protocols. (Though if the client and server are running on the same computer, a network protocol is required as well.) The Tcl codes for the client are short, but need to be integrated into other software platforms, or called by other software (e.g., Matlab Simulink) using their interface function's API. Usually, Tcl.lib needs to be linked in this process.

An Example of the Client–Server Technique

In this example, OpenSees server builds a structure model and waits for commands from a client. The client sends the command "runOpenSeesOneStep" with four parameters (as a string) to the server after the network "socket" is built. The server receives the command, runs it, calculates the resisting force of the structure's rigid foundation, sends back the results to the client, and then waits for the next command. Lines S10–S14 and line S19 need to be uncommented, and line 15 needs to be commented for testing this example.

The model.tcl in line S19 consists of an FE model and other user-defined macro commands such as "getTotalResistingForce" and "runOpenSeesOneStep". This example uses a simple reinforced concrete frame structure built on a rigid foundation, e.g., a shaking table or a rigid foundation in soil structure interaction (SSI) problems. The Tcl codes are as follows:

```
1   model BasicBuilder -ndm 2 -ndf 3
2   global Me
3   set Me [expr 7.55e6/2.0]
4   global H
5   global L
6   set H 36.3
7   set L 25.2
8   node 1 0.0 0.0
9   node 2 0.0 $H -mass $Me $Me 0.0
10  node 4 $L 0.0
11  node 3 $L $H -mass $Me $Me 0.0
12  geomTransf Linear 1
13  set Area [expr 6.1*6.1]
14  set Iz [expr 6.1*6.1*6.1*6.1/12.0]
15  set factorK 1.0
16  element elasticBeamColumn 1 1 2 $Area [expr 2.7e10*$factorK]
    $Iz 1
17  element elasticBeamColumn 2 2 3 $Area [expr 2.0e14*$factorK]
    $Iz 1
18  element elasticBeamColumn 3 3 4 $Area [expr 2.7e10*$factorK]
    $Iz 1
19  constraints Transformation
20  test NormDispIncr 1.E-6 25 1
21  algorithm Newton
22  numberer RCM
```

```
23  system BandGeneral
24  set w1 13.9817881546 ;
25  set w2 85.6248678831 ;
26  set ksi 0.02
27  set a0 [expr $ksi*2.0*$w1*$w2/($w1+$w2)]
28  set a1 [expr $ksi*2.0/($w1+$w2)]
29  integrator Newmark 0.5 0.25
30  rayleigh $a0 0.0 $a1 0.0
31  analysis Transient
32  recorder Node -file disp.out -time -node 2 3 -dof 1 2 3 disp
33  recorder Node -file reaction.out -time -node 1 4 -dof 1 2 3
    reaction
```

Lines 1–33 build the RC frame structure.

```
34  proc runOpenSeesOneStep { numOfTimeStep timeStep
    currentDisp_1 currentDisp_2} {
35  global L
36  remove loadPattern 1 ; # First delete the pattern {\#}1, then add a
    new pattern #1 below
37  pattern MultipleSupport 1 {
38  groundMotion 11 Plain -disp "Series -time { [expr
    $numOfTimeStep * $timeStep] } -values { $currentDisp_1} "
39  imposedMotion 1 1 11
40  set currentDisp_y [expr -$currentDisp_2*$L/2 ]
41  groundMotion 12 Plain -disp "Series -time { [expr
    $numOfTimeStep * $timeStep] } -values { $currentDisp_y } "
42  imposedMotion 1 2 12
43  groundMotion 13 Plain --disp "Series -time { [expr
    $numOfTimeStep * $timeStep] } -values { $currentDisp_2} "
44  imposedMotion 1 3 13
```

Commands 43 and 44 exert rotational displacement excitation on node 1.

```
45  groundMotion 14 Plain -disp "Series -time { [expr
    $numOfTimeStep * $timeStep] } -values { $currentDisp_1} "
```

```
46  imposedMotion 4  1  14
47  set currentDisp_y [expr $currentDisp_2*$L/2 ]
48  groundMotion  15  Plain -disp "Series -time { [expr
    $numOfTimeStep * $timeStep] } -values { $currentDisp_y} "
49  imposedMotion  4  2  15
50  groundMotion  16  Plain -disp "Series -time { [expr
    $numOfTimeStep * $timeStep] } -values { $currentDisp_2} "
51  imposedMotion  4  3  16
    }
52  analyze 1 [expr $timeStep]
    }
```

Lines 34–52 define a procedure "runOpenSeesOneStep" which applies the base excitation load. The four parameters are the number of steps, the time step, the horizontal displacement, and the rotational displacement of the foundation. Lines 37–51 define a MultipleSupport to apply horizontal, vertical, and rotation displacements to nodes 1 and 4. For example, line 38 defines ground motion #11, a displacement series whose value is $currentDisp_1 at a specific time ($numOfTimeStep * $timeStep). Line 39 applies ground motion #11 horizontally to node 1. The vertical displacements of nodes 1 and 4 are calculated by using the rotation multiplied by their distance to the center of the building (Lines 40 and 47). Line 52 performs one-step analysis.

> *Note*: The command "Series -time {[expr $numOfTimeStep * $timeStep]}-values {$currentDisp_1}" may cause problems. As mentioned above, if the current time of OpenSees is slightly larger than "$numOfTimeStep * $timestep", even a very small error of 10^{-16} s will cause the displacement to become zero. Therefore, the series may be defined as follows:
>
> Series -time {[expr $numOfTimeStep * $timeStep - 1.0e-10] [expr$numOfTimeStep * $timeStep + 1.0e-10]}-values {$currentDisp_1 $currentDisp_1}

The procedure "getTotalResistingForce" calculates the reaction force of the foundation (i.e., the center of nodes 1 and 4) on the superstructure as follows:

```
53  proc getTotalResistingForce {} {
54  global Me
55  global H
56  global L
57  global Fx
58  global My
59  set Fx1 [nodeReaction 1 1]
60  set Fx2 [nodeReaction 4 1]
61  set Fx [expr $Fx1+$Fx2]
62  set Fy1 [nodeReaction 1 2 ]
63  set Fy2 [nodeReaction 4 2 ]
64  set My1 [nodeReaction 1 3]
65  set My2 [nodeReaction 4 3]
66  set My [expr $Fy1*$L/2-$Fy2*$L/2-$My1-$My2]
```

Lines 61–66 are the reactions of the foundation (the horizontal force and moment), which are stored in the global variables "F_x" and "M_y". Vertical force is not considered in this example (e.g., for SSI problems).

Note: When using the Tcl command "nodeReaction 1 1" (e.g., line 59), a recorder must be used in the model to record the reaction (see line 33), otherwise the command "nodeReaction" will return a wrong result.

The analysis results are presented in the Fig. 10.3.

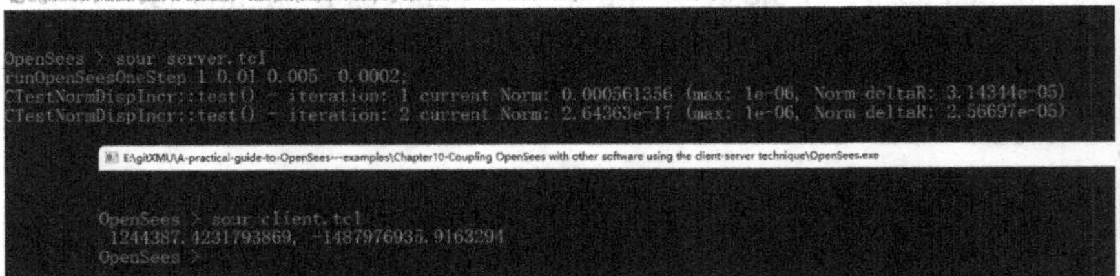

Fig. 10.3 Analysis results using the CS technique.

Reference

[1] Gu, Q. and Ozcelik, O. Integrating OpenSees with other software-with application to coupling problems in civil engineering. *Struct. Eng. Mech. Int. J.* 2011, *40*(1), 85–103.

Chapter 11

Response Sensitivity Analysis Based on Direct Differentiation Method (DDM)

The direct differentiation method (DDM) is an efficient and accurate method to compute the response sensitivities [1]. In this chapter a simple model with two truss elements shown in Fig. 11.1 is used to perform sensitivity analysis using DDM. The material type is "Steel01", whose parameters are yield strength, initial elastic modulus, and strain hardening ratio. The displacement sensitivity with respect to the elastic modulus (E) is calculated using the DDM. A forward finite difference (FFD) method will be applied to verify the DDM results. Both static and transient analyzes will be performed. It is worth mentioning that the DDM used to work well until recently (due to some newly updated codes that are incompatible with the DDM framework). The authors provide a workable source code at https://github.com/OpenSeesXMU.

The Tcl commands are as follows:

```
1   wipe
2   model basic -ndm 2 -ndf 2
3   reliability
```

Line 3 creates a reliability domain, which is necessary for sensitivity analysis in OpenSees.

```
4   node 1    0.0  0.0
5   node 2   10.0  0.0
6   node 3   20.0  0.0
```

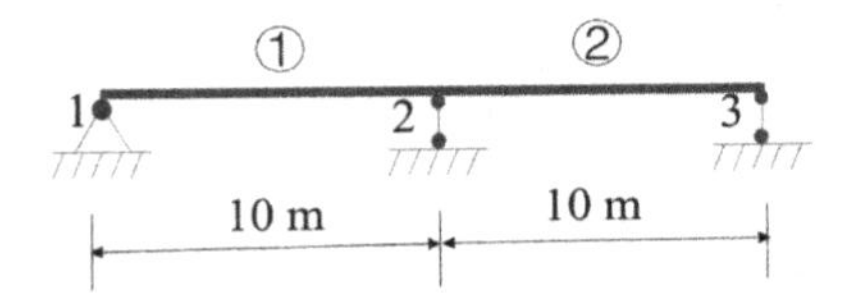

Fig. 11.1 Model of the example.

```
7   fix 1 1 1
8   fix 2 0 1
9   fix 3 0 1
10  uniaxialMaterial Steel01 1 248200 2.0e8 0.05
11  element truss  1  1  2  0.1  1
12  element truss  2  2  3  0.1  1
13  parameter 1 -element 1 -material E
14  addToParameter 1 -element 2 -material E
```

Lines 13 and 14 define sensitivity parameters. In line 13, "parameter 1" defines a sensitivity parameter tagged 1, "-element 1 -material E" maps the material parameter E (the initial elastic modulus) of element #1 to sensitivity parameter #1. Since the elastic modulus of both elements are sensitivity parameters, "addToParameter" is used in line 14 to map the E of element #2 to parameter #1 also.

```
15  recorder Node -file ddm2E.out -time -precision 16 -node 2 -dof 1
      "sensitivity 1"
```

In line 15, "sensitivity 1" — the displacement sensitivity with respect to parameter E — is recorded. "1" is the tag of the sensitivity parameter.

```
16  recorder Node -file node2.out -time -precision 16 -node 2 -dof 1
      2 disp
17  recorder Element -file force.out -time -precision 16 -ele 1
      axialForce
18  recorder Element -file deformation.out -time -precision 16 -ele 1
      deformations
19  timeSeries Path 1 -time {0 0.5 1.0 10000.0} -values {0 0.5
      1.0 1.0}
20  pattern Plain 1 1 {
```

```
21  load 3 2e4 0.0
22  }
23  constraints Transformation
24  test NormDispIncr 1e-12 25 2
25  algorithm Newton
26  numberer RCM
27  system BandGeneral
28  integrator LoadControl 0.1;
29  analysis Static
30  sensitivityAlgorithm -computeAtEachStep
31  analyze 10
```

Line 30 indicates that the response sensitivity will be calculated at each time step.

```
32  wipeAnalysis
33  pattern UniformExcitation 2 1 -accel "Series -factor 300
    -filePath el.txt -dt 0.01"
34  mass 1 100.0 0.0
35  mass 2 100.0 0.0
36  mass 3 100.0 0.0
37  constraints Transformation
38  test NormDispIncr 1e-12 25 2
39  algorithm Newton
40  numberer RCM
41  system BandGeneral
42  integrator Newmark 0.5 0.25
43  analysis Transient
44  sensitivityAlgorithm -computeAtEachStep
45  analyze 100 0.01
```

Line 44 is the same as line 30, indicating that the response sensitivity will be calculated at each time step.

Analysis results: The displacement sensitivity with respect to parameter E is shown in Fig. 11.2. In the figure, "u" and "θ" represent responses and variables, respectively, and they are displacement and E for Fig. 11.2. The DDM result and the FFD results with varying perturbations are compared.

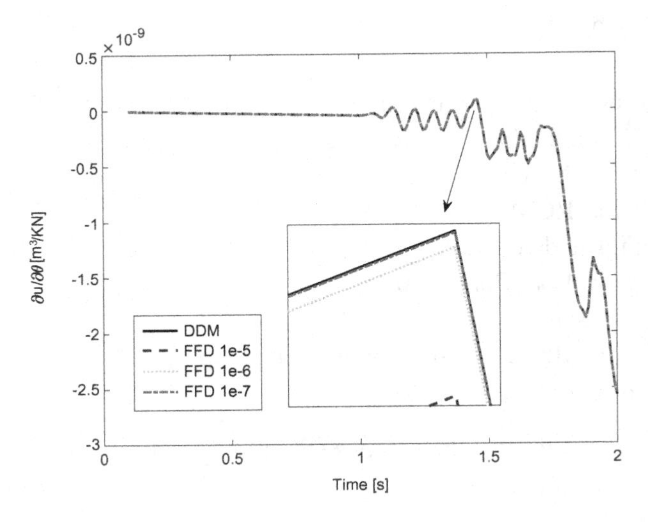

Fig. 11.2 Comparison of displacement sensitivities with respect to parameter E using the DDM and FFD.

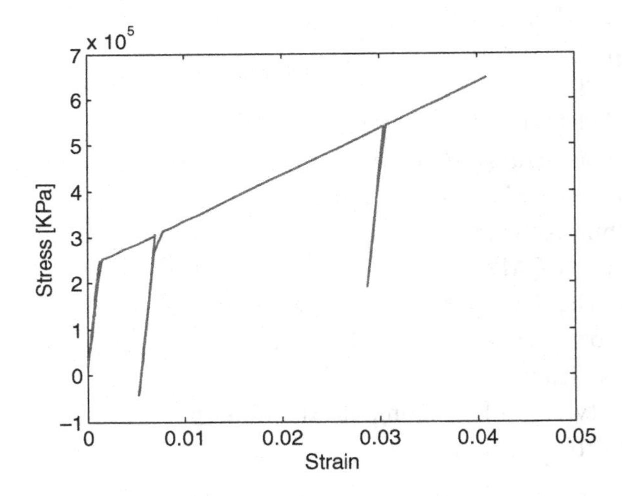

Fig. 11.3 The constitutive behavior of steel01 material in element #1.

The FFD results converge to that of DDM with decreasing perturbation, verifying the DDM algorithm. The constitutive behavior of element #1 is shown in Fig. 11.3, which indicates that the model has experienced nonlinearities.

Reference

[1] Gu, Q., Conte, J. P., Elgamal, A., *et al.* Finite element response sensitivity analysis of multi-yield-surface J2 plasticity model by direct differentiation method. *Comput. Methods Appl. Mech. Eng.* 2009, *198* (30–32): 2272–2285.

Chapter 12

Structural Analysis Based on Bond-Based Peridynamics (BPD)

12.1 Introduction to BPD Method

This chapter introduces how to study nonlinear structural behaviors by using bond-based peridynamics (BPD) theory in OpenSees. Peridynamics (PD) is a type of mesh-free method proposed by Silling and co-workers that is a reformulation of continuum mechanics. PD has the natural advantage of analyzing discontinuous problem of structures with cracks and fractures. PD includes BPD and state-based peridynamics (SPD). This section introduces BPD, while SPD will be introduced in Chapter 18.

According to the BPD theory, a physical domain is discretized into PD particles. Each particle is assumed to be subjected to the forces from all particles inside its horizon, e.g., a spherical neighborhood of the particle. The equation of motion at point **x** and time t is given as Eq. (12.1) where a two-particle force function **f** is used to describe the interaction between material particles, as shown in Fig. 12.1.

$$\rho(\mathbf{x})\ddot{\mathbf{u}}(\mathbf{x},t) = \int_{\mathscr{H}_\mathbf{x}} \mathbf{f}\left(\mathbf{u}\left(\mathbf{x}',t\right) - \mathbf{u}(\mathbf{x},t), \mathbf{x}' - \mathbf{x}\right) dV_{\mathbf{x}'} + \mathbf{b}(\mathbf{x},t) \qquad (12.1)$$

where $\mathscr{H}_\mathbf{x}$ denotes the horizon of particle **x**. **u** is a displacement vector field and $\ddot{\mathbf{u}}$ is acceleration. **b** is body force density, ρ is mass density in the reference configuration, and **f** is a pairwise force function whose value is the force vector (per unit volume) that particle $\mathbf{x}'$ exerts on particle **x**. Each force vector **f** (per unit volume) herein is assumed to be a function (linear of nonlinear) of elongation between particle $\mathbf{x}'$ and particle **x**. Practically, the

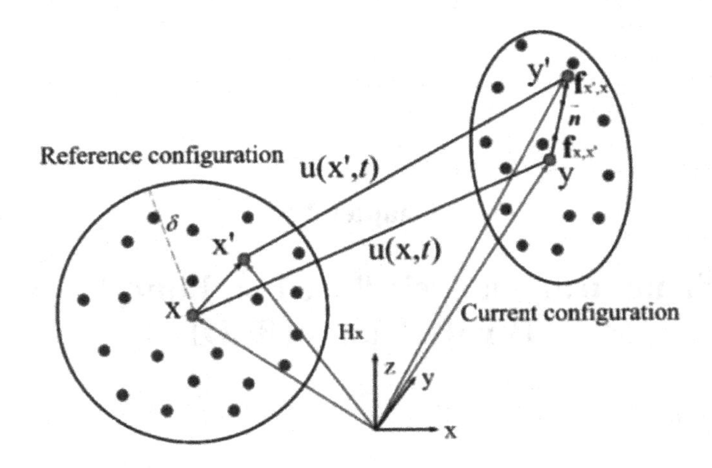

Fig. 12.1 Terminology for a peridynamic model.

first term in the right-hand side of equation (12.1) can be calculated as

$$\mathbf{f}(\boldsymbol{\eta},\boldsymbol{\xi}) = \sum_{i=1}^{\text{nel}} [\mathrm{f}_i(s)]\, \boldsymbol{n}_i \tag{12.2}$$

where f_i is the force of the ith bond, i.e., between particle **x** and another point, and nel is the number of bond (i.e, all points within the horizon of **x**). $\boldsymbol{n}_i$ is a direction along the bond. $\boldsymbol{\eta}+\boldsymbol{\xi}$ represents the connecting direction of the two particles of the ith bond in the current configuration. s is the bond stretch, which is defined as

$$s = \frac{|\boldsymbol{\eta}+\boldsymbol{\xi}|-|\boldsymbol{\xi}|}{|\boldsymbol{\xi}|} \tag{12.3}$$

For nonlinear materials, the initial elastic modulus of the bond is defined as $E = \left.\frac{\partial f(s)}{\partial s}\right|_{s=0}$, which is calculated using the following equations (proposed by Silling [1]):

$$\begin{aligned} E &= \frac{9E_{\text{macro}}}{\pi h \delta^3} && \text{for 2D model} \\ E &= \frac{6E_{\text{macro}}}{\pi \delta^4 (1-2v)} && \text{for 3D model} \end{aligned} \tag{12.4}$$

where E_{macro} is the macro elastic modulus and v is Poisson's ratio.

The equilibrium equation of BPD theory in Eq. (12.1) can be rewritten as another format, i.e., an equilibrium equation being the same as that of FEM in Eq. (15.1). Equation (15.1) can be discretized and solved in the same manner with FEM, e.g., discretized along time by an implicit time stepping method and possibly solved by Newton Raphson solution algorithm. The detail can be found in Section 15.3.

12.2 Application Example of BPD Analysis in OpenSees

To apply the BPD modeling method in OpenSees, the PD particle is defined by node (same as FE node), and the force vector **f** is defined by the truss element of FEM. The physical meaning of bond stretch is the same as strain in FEM. The elastic modulus E is regarded in peridynamics as the material's stress–strain relationship. In the PD model in this example, a central-difference explicit integration is utilized to solve the motion equation. In this section, the BPD model is used to simulate the tensile failure (e.g., crack propagation) of a slab with pre-crack. The BPD modeling of the board is shown in Fig. 12.2.

The analysis and modeling code of this example is as follows:

```
1   model basic -ndm 2 -ndf 2
2   set nx 40;set dx 1.0;set horizon 1.5;
3   source nodebuild.tcl
4   nodebuild $nx $nx $dx
5   uniaxialMaterial Elastic 1 2.0e11
6   source elementbuild.tcl
```

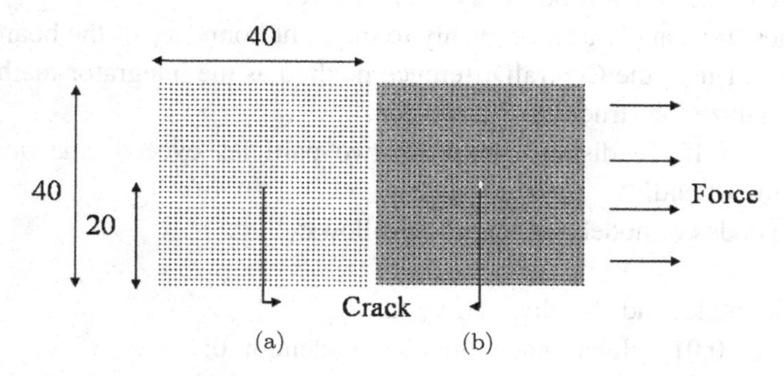

Fig. 12.2 The BPD model of a slab with pre-crack (a) PD particles; (b) particles and bonds.

```
7   elementbuild $nodenum [expr $horizon*$dx] $nx
8   fixX 0.0 1 1
9   set timestep 0.000000025
10  source timeSeries.tcl
11  pattern MultipleSupport 1 {
12  groundMotion 11 Plain -disp 22
13  for { set i [expr $nodenum-$nx+1] } { $i <= $nodenum} { incr i}
    {
14      imposedMotion $i 1 11
15  }}
16  constraints Transformation
17  numberer Plain
18  system BandGeneral
19  test NormDispIncr 1.0e-8 6 2
20  algorithm Linear
21  integrator CentralDifference
22  analysis Transient
23  for { set times 1 } { $times <= 6000} { incr times} {
24  analyze 1 $timestep
25  source remove.tcl }
```

Line 2 sets a slab with a size of 40 * 40, the grid spacing is 1.0, and the horizon is 1.5.

Lines 3–4 create the PD model by uniformly discretizing the domain into PD particles.

Line 5 defines the material of the bond.

Lines 6–7 create all the PD bonds.

Line 8 fixes the left boundary of the board.

Lines 9–15 apply a force evenly to the right boundary of the board.

Line 21 uses the CentralDifference method as the integrator method to static analyze the structure.

Line 25 If the distance between two particles exceeds the horizon, delete their bond.

The codes of nodebuild.tcl are as follows:

```
1   proc nodebuild { ndivx ndivy dx} {
2   set m 0.01; global nodenum; set nodenum 0;
3   for { set i 1 } { $i <= $ndivx} { incr i} {
4   for { set j 1 } { $j <= $ndivy} { incr j} {
```

```
5  set nodenum [expr $nodenum+1]
6  node $nodenum [expr ($i-1)*$dx] [expr ($j-1)*$dx] -mass  $m
   $m
7  }}}
```

The codes of elementbuild.tcl are as follows:

```
1  proc elementbuild { nnum horizon ndivx} {
2  global elenum; global active; global bond; set elenum 0;
3  for { set i 1 } { $i <= [expr $nnum-1]} { incr i} {
4  if { [expr $i+$ndivx+int($horizon)+1]<= $nnum} {
5  set maxnode [expr $i+$ndivx+int($horizon)+1]
6  } else { set maxnode $nnum}
7  or { set j [expr $i+1] } { $j <= $maxnode} { incr j} {
8  if {761<=$i && $i<=781 && 801<=$j && $j<=821} {} else {
9  set cAx [nodeCoord $i 1];set cAy [nodeCoord $i 2];
10 set cBx [nodeCoord $j 1];set cBy [nodeCoord $j 2];
11 set length [expr pow(pow(($cBx-$cAx),2) + pow(($cBy-
   $cAy),2),0.5)];
12 if {$length <= $horizon} {
13 set elenum [expr $elenum+1]
14 element truss $elenum  $i  $j  1.0 1
15 set bond($elenum,1) $i;set bond($elenum,2) $j;
16 set active($elenum,1) 1
17 }}}}}
```

The codes of remove.tcl are as follows:

```
1  for { set i 2984 } { $i <= 3080} { incr i} {
2  if {$active($i,1)==1} {
3  set cAx [expr [nodeCoord $bond($i,1) 1]+[nodeDisp $bond($i,1)
   1]];
4  set cAy [expr [nodeCoord $bond($i,1) 2]+[nodeDisp $bond($i,1)
   2]];
5  set cBx [expr [nodeCoord $bond($i,2) 1]+[nodeDisp $bond($i,2)
   1]];
6  set cBy [expr [nodeCoord $bond($i,2) 2]+[nodeDisp $bond($i,2)
   2]];
```

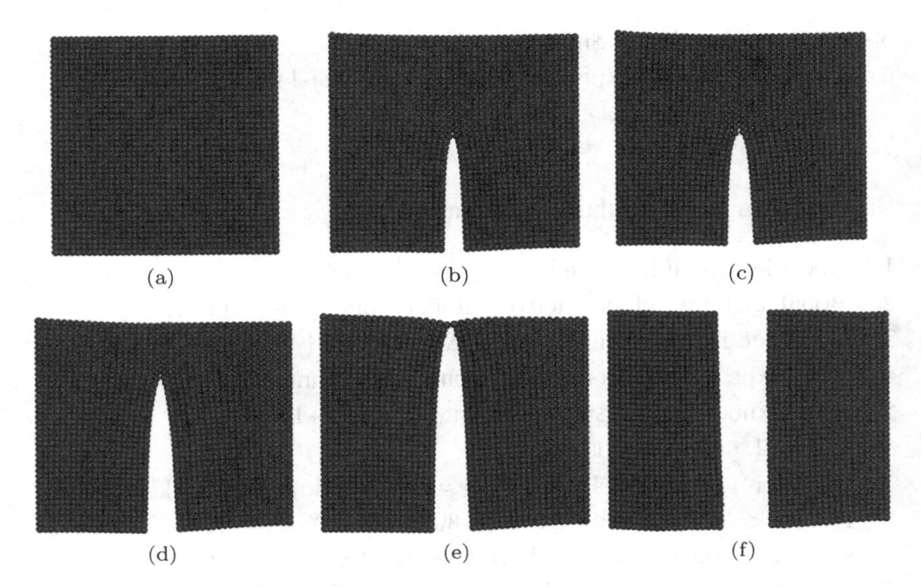

Fig. 12.3 The deformation of BPD model when displacement of right boundary is (a) 3.5; (b) 4.5; (c) 5.5; (d) 6.5; (e) 7.5; and (f) 10.0.

```
7  set length [expr pow(pow(($cBx-$cAx),2)+pow(($cBy-
   $cAy),2),0.5)]
8  if {$length>1.5} {
9  remove element $i; set active($i,1) 0 } }}
```

The deformations of the BPD model are shown in Fig. 12.3. with changing displacements of the right boundary.

12.3 Static Analysis Based on the BPD Model

The BPD has been further modified to make it applicable to perform implicit or static analysis. Similar to traditional BDP, the structures are uniformly discretized into particles each modeled using an FE node in OpenSees, named as PD particle herein. However, an implicit integration method, e.g., Newmark-βmethod, is employed to replace the explicit integration to improve the accuracy of the calculation results.

The application example is a 150 mm × 150 mm × 300 mm block of C30 concrete subjected to uniform uniaxial loading from above, as shown

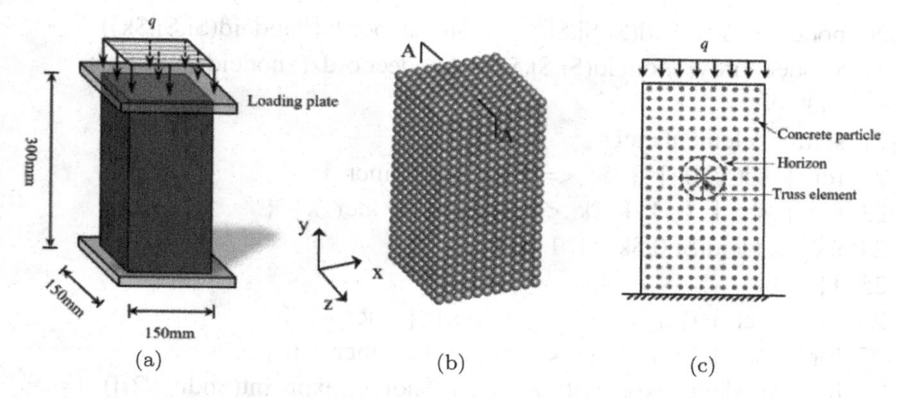

Fig. 12.4 PD modeling of concrete specimen strength test (a) overall arrangement; (b) axonometric view of the BPD model; (c) front view at section A-A.

in Fig. 12.4. The BPD modeling of the concrete specimen is shown in Figs. 12.4(b) and 12.4(c). The details of the example can be found in [2].

The analysis and modeling code of this example is as follows:

```
1  wipe
2  model basic -ndm 3 -ndf 3
3  set x 0.15; set y 0.30; set z 0.15;
4  set dx 0.015; set dy 0.015;set dz 0.015;
5  set A [expr $dx*$dx];
6  set cfpc -14.0e6;set cepsc0 -0.0014;set cfpcu -2.5e6
7  set cepsU -0.01;set clambda 0.1;set cft 2.8e6;set cEts 2e8
8  set horizonrate 2.015;set horizonratem 3.0;
9  set horizon [expr $horizonrate*$dx];set radij [expr 0.5*$dx]
10 set ndivx [expr int($x/$dx)];set ndivy [expr int($y/$dx)];
11 set ndivz [expr int($z/$dx)];set nodenum 0;
12 for { set i 0 } { $i <= $ndivx} { incr i} {
13 for { set j 0 } { $j <= $ndivy} { incr j} {
14 for { set k 0 } { $k <= $ndivz} { incr k} {
15 set nodenum [expr $nodenum+1];set nodeid($i,$j,$k) $nodenum;
16 set nodecoordx($nodeid($i,$j,$k)) [expr $dx*($i)];
17 set nodecoordy($nodeid($i,$j,$k)) [expr $dy*($j)];
18 set nodecoordz($nodeid($i,$j,$k)) [expr $dz*($k)];
19 set tag($nodeid($i,$j,$k)) 1
```

```
20 node    $nodeid($i,$j,$k)    $nodecoordx($nodeid($i,$j,$k))
   $nodecoordy($nodeid($i,$j,$k)) $nodecoordz($nodeid($i,$j,$k))
   }}}
21 source elements.tcl
22 for { set i 0 } { $i <= $ndivx} { incr i} {
23 for { set k 0 } { $k <= $ndivz} { incr k} {
24 fix $nodeid($i,0,$k) 1 1 1
25 }}
26 for { set i 0 } { $i <= $ndivx} { incr i} {
27 for { set k 0 } { $k <= $ndivz} { incr k} {
28 if { $nodeid([expr int($ndivx/2)],$ndivy,[expr int($ndivz/2)]) !=
   $nodeid($i,$ndivy,$k) } {
29 equalDOF    $nodeid([expr    int($ndivx/2)],$ndivy,[expr
   int($ndivz/2)]) $nodeid($i,$ndivy,$k) 2           }}}
30 for { set i 0 } { $i <= $ndivx} { incr i} {
31   for { set k 0 } { $k <= $ndivz} { incr k} {
32         fix $nodeid($i,$ndivy,$k) 1  0 1
33           }}
34 pattern Plain 1 Linear {
35 for { set i 0 } { $i <= $ndivx} { incr i} {
36   for { set k 0 } { $k <= $ndivz} { incr k} {
37         load $nodeid($i,$ndivy,$k) 0.0 -2.0e6 0.0
38 }}}
39 set startT [clock seconds]
40 constraints Transformation
41 test NormDispIncr 2.0e-5 15 1
42 algorithm Newton
43 system Mumps
44 numberer Plain
45 integrator    DisplacementControl    $nodeid([expr
   $ndivx/2],$ndivy,[expr $ndivz/2]) 2 -0.000005
46 analysis Static
47 analyze 200
48 set endT [clock seconds]
49 puts "Over time: [expr $endT-$startT] seconds."
```

Line 2 defines a 3D model and three degrees of freedom.

Line 3 sets the size of concrete column, 0.15 m × 0.30 m × 0.15 m.

Line 4 sets mesh size to 0.015 m.

Lines 6–7 set the parameters of uniaxial material Concrete02 used for truss element.

Lines 8–9 set the radius length of the horizon.

Lines 10–20 create the PD model by uniformly discretizing the domain into PD particles.

Line 21 searches for and builds the bond relationships within its horizon for each PD particle.

Lines 22–25 fix the degrees of freedom in three directions at the bottom particles of the concrete column.

Lines 26–33 simulate the impact of the loading plate

Lines 34–38 apply vertical load on the top of the concrete column

Lines 39–47 set the analysis method and iterative control approach, and so on.

The codes of element.tcl are as follows:

```
1  for { set i 0 } { $i <= $ndivx} { incr i} {
2  for { set j 0 } { $j <= $ndivy} { incr j} {
3  for { set k 0 } { $k <= $ndivz} { incr k} {
4  set imax [expr int($i+$horizonratem)];set imin [expr int($i-
   $horizonratem)];set jmax [expr int($j+$horizonratem)];set jmin
   [expr int($j-$horizonratem)];set kmax [expr
   int($k+$horizonratem)];set kmin [expr int($k-$horizonratem)];
5  if {$imax>$ndivx} {
6  set imax [expr $ndivx]}
7  if {$jmax>$ndivy} {
8  set jmax [expr $ndivy] }
9  if {$kmax>$ndivz} {
10 set kmax [expr $ndivz] }
11 if {$imin<0} {
12 set imin [expr int(0)]  }
13 if {$jmin<0} {
14 set jmin [expr int(0)]  }
15 if {$kmin<0} {
```

```
16 set kmin [expr int(0)]   }
17 set cellnum($nodeid($i,$j,$k)) 0
18  for { set l $imin } { $l <= $imax} { incr l} {
19        for { set m $jmin } { $m <= $jmax} { incr m} {
20         for { set n $kmin } { $n <= $kmax} { incr n} {
21                 if {$nodeid($i,$j,$k)<$nodeid($l,$m,$n)} {
22        set Cid $nodeid($i,$j,$k); set Tid $nodeid($l,$m,$n);
23 } else {
24        set Tid $nodeid($i,$j,$k);set Cid $nodeid($l,$m,$n);
25 }
26                    set contact($Cid,$Tid) 0
27       }}}}}}
28 set trussnum 0
29 for { set i 0 } { $i <= $ndivx} { incr i} {
30  for { set j 0 } { $j <= $ndivy} { incr j} {
31   for { set k 0 } { $k <= $ndivz} { incr k} {
32 set imax [expr int($i+$horizonratem)]; set imin [expr int($i-
   $horizonratem)];set jmax [expr int($j+$horizonratem)]; set jmin
   [expr int($j-$horizonratem)];set kmax [expr
   int($k+$horizonratem)];set kmin [expr int($k-$horizonratem)];
33 if {$imax>$ndivx} {
34 set imax [expr $ndivx]   }
35 if {$jmax>$ndivy} {
36 set jmax [expr $ndivy]   }
37 if {$kmax>$ndivz} {
38 set kmax [expr $ndivz]   }
39 if {$imin<0} {
40 set imin [expr int(0)]
41    }
42 if {$jmin<0} {
43 set jmin [expr int(0)]   }
44 if {$kmin<0} {
45 set kmin [expr int(0)]   }
46 set cellnum($nodeid($i,$j,$k)) 0
```

```
47  for { set l $imin } { $l <= $imax} { incr l} {
48      for { set m $jmin } { $m <= $jmax} { incr m} {
49       for { set n $kmin } { $n <= $kmax} { incr n} {
50             if {$nodeid($i,$j,$k)<$nodeid($l,$m,$n)} {
51      set Cid $nodeid($i,$j,$k);set Tid $nodeid($l,$m,$n)
52  } else {
53             set Tid $nodeid($i,$j,$k);set Cid $nodeid($l,$m,$n)
    }
54              if {$contact($Cid,$Tid)==0} {
55 set          idist2($nodeid($i,$j,$k),$nodeid($l,$m,$n))       [expr
   pow($nodecoordx($nodeid($i,$j,$k))-
   $nodecoordx($nodeid($l,$m,$n)),2)+pow($nodecoordy($nodeid(
   $i,$j,$k))-
   $nodecoordy($nodeid($l,$m,$n)),2)+pow($nodecoordz($nodeid(
   $i,$j,$k))-$nodecoordz($nodeid($l,$m,$n)),2)]
56 set          idist($nodeid($i,$j,$k),$nodeid($l,$m,$n))       [expr
   pow($idist2($nodeid($i,$j,$k),$nodeid($l,$m,$n)),0.5)]
57 if      {$idist($nodeid($i,$j,$k),$nodeid($l,$m,$n))>0.0       &&
   $idist($nodeid($i,$j,$k),$nodeid($l,$m,$n))<= [expr $horizon-
   $radij]} {
58 set trussnum [expr $trussnum+1]
59 set family($trussnum,1) $nodeid($i,$j,$k);
60 set family($trussnum,2) $nodeid($l,$m,$n)
61 set fac($trussnum) 1.0
62                            if      {$tag($nodeid($i,$j,$k))==1
&&$tag($nodeid($l,$m,$n))==1}     {
63 uniaxialMaterial Concrete02 $trussnum $cfpc $cepsc0 $cfpcu
   $cepsU $clambda $cft $cEts
64 element truss $trussnum $nodeid($i,$j,$k) $nodeid($l,$m,$n)
   $A    $trussnum
65 set family($trussnum,3) 1      }
66 set          nodelink([expr         int($nodeid($i,$j,$k))],[expr
   int($nodeid($l,$m,$n))])      $trussnum
67 set             cellnum($nodeid($i,$j,$k))                 [expr
   $cellnum($nodeid($i,$j,$k))+1]
```

```
68 set cell($nodeid($i,$j,$k),$cellnum($nodeid($i,$j,$k)))
   $nodeid($l,$m,$n)
69 if {$idist($nodeid($i,$j,$k),$nodeid($l,$m,$n))>[expr $horizon-
   $radij] && $idist($nodeid($i,$j,$k),$nodeid($l,$m,$n))< [expr
   $horizon+$radij]} {
70 set trussnum [expr $trussnum+1]
71 set family($trussnum,1) $nodeid($i,$j,$k)
72 set family($trussnum,2) $nodeid($l,$m,$n)
73 set fac($trussnum) [expr ($horizon+$radij-
   $idist($nodeid($i,$j,$k),$nodeid($l,$m,$n)))/(2.0*$radij)]
74 if {$tag($nodeid($i,$j,$k))==1 && $tag($nodeid($l,$m,$n))==1}
   {
75 uniaxialMaterial Concrete02 $trussnum [expr
   $cfpc*$fac($trussnum)] $cepsc0 [expr $cfpcu*$fac($trussnum)]
   $cepsU $clambda [expr $cft*$fac($trussnum)] [expr
   $cEts*$fac($trussnum)]
76 element truss $trussnum $nodeid($i,$j,$k)
   $nodeid($l,$m,$n) $A $trussnum
77 set family($trussnum,3) 1; set contact($Cid,$Tid)
78 }
79 set nodelink([expr int($nodeid($i,$j,$k))],[expr
   int($nodeid($l,$m,$n))]) $trussnum
80 set cellnum($nodeid($i,$j,$k)) [expr
   $cellnum($nodeid($i,$j,$k))+1]
81 set cell($nodeid($i,$j,$k),$cellnum($nodeid($i,$j,$k)))
   $nodeid($l,$m,$n) }}}}}}}}
```

The simulation results are compared with the test results as shown in Fig. 12.5. It is observed that the BPD modeling can capture the main features of the plastic behaviors of concrete specimens in the monotonic loading cases, e.g., the strength deterioration.

The corresponding 3D diagrams of concrete specimens are shown in Fig. 12.6 where the displacements are scaled up 50 times to show the cracks.

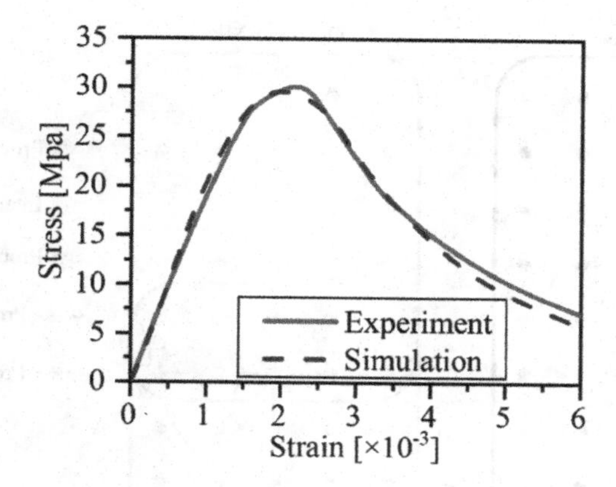

Fig. 12.5 The stress–strain relationships of concrete in uniaxial strength test under compressive loading.

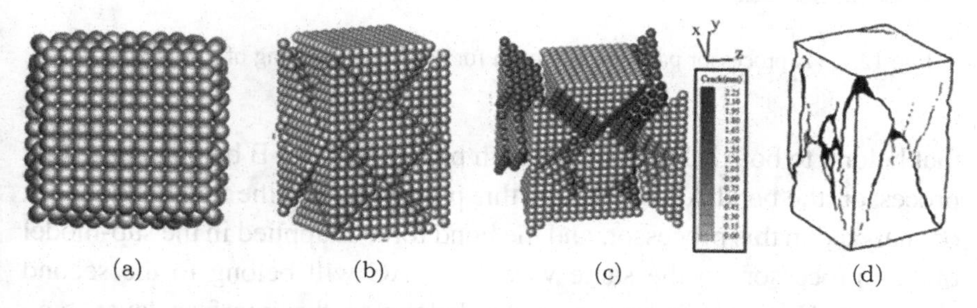

Fig. 12.6 The deformation of the concrete (a) top view in stage I, (b) stage II, (c) stage III, and (d) crack patterns of the specimens with high friction boundary conditions.

12.4 Parallel Computing of the BPD Model

A BPD model may have a large amount of particles and bonds, and the computational cost is usually expensive. Parallel computing may be used to reduce the computing costs. A parallel computing approach in OpenSeesMP is proposed herein to analyze the enhanced BPD model. Figure 12.7 shows an example of dividing the original BPD model into two sub-models each of which is analyzed using an individual processor. The two processors interact through the so-called "interactive particles"

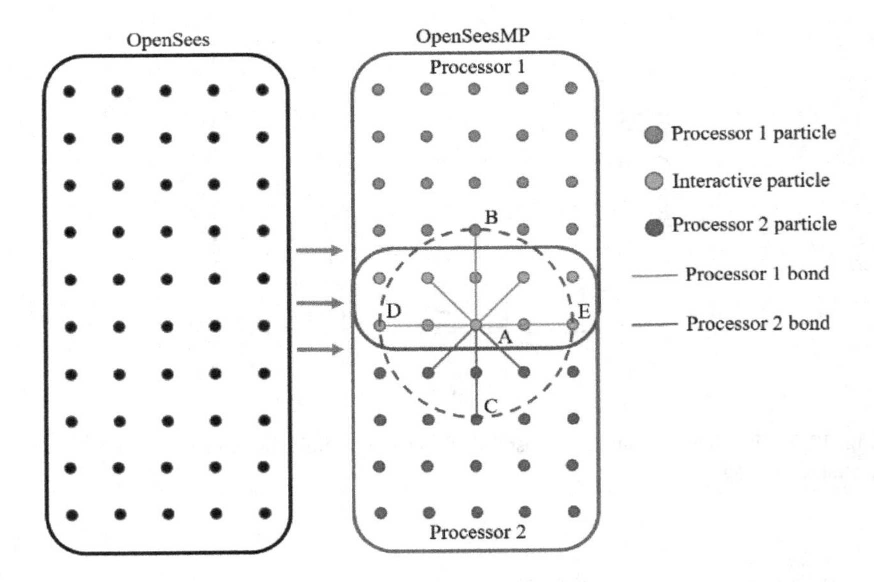

Fig. 12.7 A processor partition approach for parallel computing of the BPD model.

that belong to both sub-models. If both particles A and B belong to the first processor, the bond AB belongs to this processor, i.e., the bond AB is built or analyzed in this processor, and the bond force is applied in the sub-model of this processor. In the same way, bond AC will belong to the second processor. However, if both particles belong to the interface layer, e.g., bond AD or AE, the bonds can belong to either sub-model (user decides), but cannot be repeated. The details of the parallel computing approach for the BPD model can be found in [3].

Example in Section 12.3 has been rewritten into parallel form according to the form of OpenseesMP, as follows:

```
1    set pid [getPID];set np [getNP]
2    if {$np != 2} exit
3    model BasicBuilder -ndm 3 -ndf 3
4    source material.tcl
5    if {$pid == 0} {
6    source node0.tcl
7    source fix.tcl
```

```
8   source element0.tcl
9   } else {
10  source node1.tcl
11  source element1.tcl
12  source force.tcl
13  }
14  set startT [clock seconds]
15  constraints Transformation
16  test NormDispIncr 2.0e-5 15 1
17  algorithm Newton
18  system Mumps
19  numberer ParallelPlain
20  integrator DisplacementControl $nodeid([expr
    $ndivx/2],$ndivy,[expr $ndivz/2]) 2 -0.000005
21  analysis Static
22  analyze 200
23  set endT [clock seconds]
24  puts "Over time: [expr $endT-$startT] seconds."
```

Different from the codes in Chapter 12.3

Line 1 returns the number of processes in computation (two in this example).

Line 2 returns unique process id {0,1, .. NP-1}

Line 3 confirms that the example has only two processors

Line 5 defines the properties of the uniaxial material.

Lines 6–9 define the nodes that belong to process 1 and interface, define the elements that belong to process 1, and define the fix that belongs to process 1.

Lines 10–13 define the nodes that belong to process 2 and interface, define the elements that belong to process 2, and define the load that belongs to process 2.

Line 20 uses ParallelPlain to replace Plain

The result obtained by OpenSeesMP is perfectly consistent with that by OpenSees, as shown in Fig. 12.8. In this example, the calculation time using two processors is half of that using a single processor (considering only the analysis time).

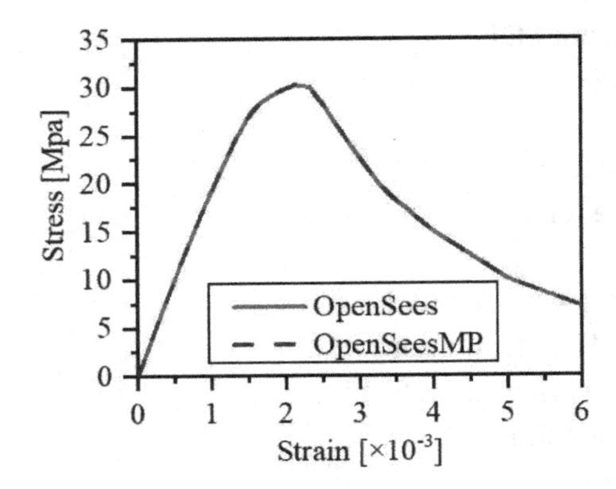

Fig. 12.8 The comparative results between the OpenSees and OpenSeesMP.

References

[1] Silling, S. Reformulation of elasticity theory for discontinuities and long-range forces. *J. Mech. Phys. Solids*, 2000, *48*(1): 175–209.

[2] Zhang, N. and Gu, Q. A practical bond-based peridynamic modeling of reinforced concrete structures. *Eng. Struct.* 2021, *244*: 112748.

[3] Zhang, N. and Gu, Q. Refined peridynamic modelling of bond-slip behaviors between ribbed steel rebar and concrete in pull-out test. *J. Struct. Eng.* 2022, 148(12): 04022197.

Chapter 13

A Wheel–Rail Interaction Element for High-Speed Railway System

13.1 A Two-Dimensional (2D) Wheel–Rail Interaction (WRI) Element

This section introduces a practical 2D WRI element [1] for simulating vertical WRI in vehicle-track-bridge (VTB) coupling systems. A typical VTB system includes subsystems of vehicles and rail-bridge as shown in Fig. 13.1. The WRI element simulates the time- and location-varying WRI using its internal resisting force, enabling the finite element (FE) model of the VTB system to remain unchanged in analysis (to be opposite to the iterative procedure between the two subsystems). This element is capable of simulating the nonlinear WRI behaviors and the rail irregularity. Each WRI element includes a wheel node and all the nodes of rail elements that the wheel may potentially pass over during the movement. For example, there are four WRI elements for a vehicle, as shown in Fig. 13.1. Details can be found in the literature [1].

The following is an example of dynamic analysis of a simple VTB system where the middle portion is a bridge of length 30 m. As shown in Fig. 13.2, the vehicle consists of a car body, two bogies, and four wheels, which are simulated by nodes with mass, Euler–Bernoulli beam elements, and WRI elements, and the connections between them are simulated by truss elements. The velocity of the train is 100 km/h.

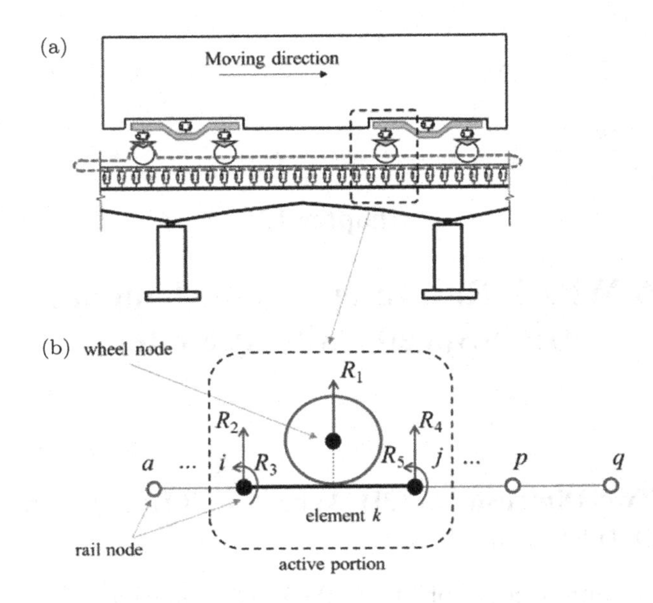

Fig. 13.1 A VTB system and a 2D WRI element.

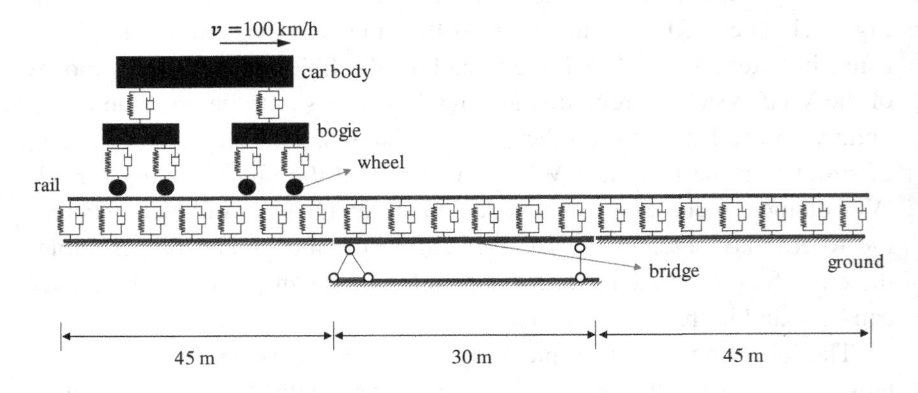

Fig. 13.2 FE model of VTB system.

The Tcl commands for this example are as follows:

```
1. wipe
2. model basic -ndm 2 -ndf 3
3. set      pInitLocation       10
```

```
4. set pDeltT 0.001
5. set pVel 27.78
6. set pRWheel 0.4575
7. set pI 4.074e-5
8. set pE 2.06e11
9. set pA 77.45e-4
10. set transfTag 1
11. geomTransf Linear $transfTag
```

The above lines define the basic information of the model by various variables. Lines 3–5 define the initial position of the vehicle, size of the time step, and the velocity of the vehicle, respectively. Lines 6–9 define the parameters of wheel and rail. Then the bridge and train models are defined using two separate files, i.e., bridge.tcl and train.tcl.

```
12. source bridge.tcl
13. source train.tcl
```

The details of the two models in line 12 and 13 will be introduced later. Then WRI for the four wheels are defined by using four WRI elements, i.e.,

```
14. set pInitLocation1 [expr 0+$pInitLocation]
15. set pInitLocation2 [expr 2.5+$pInitLocation]
16. set pInitLocation3 [expr 17.5+$pInitLocation]
17. set pInitLocation4 [expr 20+$pInitLocation]
18. element WheelRail 3001 $pDeltT $pVel $pInitLocation1 2001
    $pRWheel $pI $pE $pA $transfTag 10 -NodeList $NodeID
    rail_Irreg.txt;
19. element WheelRail 3002 $pDeltT $pVel $pInitLocation2 2002
    $pRWheel $pI $pE $pA $transfTag 10 -NodeList $NodeID
    rail_Irreg.txt;
20. element WheelRail 3003 $pDeltT $pVel $pInitLocation3 2003
    $pRWheel $pI $pE $pA $transfTag 10 -NodeList $NodeID
    rail_Irreg.txt;
21. element WheelRail 3004 $pDeltT $pVel $pInitLocation4 2004
    $pRWheel $pI $pE $pA $transfTag 10 -NodeList $NodeID
    rail_Irreg.txt;
```

Lines 14–17 define the initial location of the four wheels. Lines 18–21 define the 2D WRI elements mentioned above. Take line 18 as an example, 3001 denotes the tag of the element. pDeltT, pVel, and $pInitLocation1 denote the size of the time step, velocity of the train, and the initial location of the wheel. 2001 denotes the wheel node tag, which can be found in train.tcl. pRWheel, pI, pE, and pA denote the radius, Moment of inertia, Young's modulus, and the sectional area of the rail, respectively. transfTag denotes the coordinate transformation. The number 10 denotes the gravity loading is applied in 10 steps of static analysis before dynamic analysis. NodeID denotes a sequence of all rail nodes that the wheel may pass through. rail_Irreg.txt refers to a file that records track irregularity.

```
22. constraints      Plain;
23. numberer         Plain;
24. system           BandGeneral;
25. test      NormDispIncr    1.0e-8  100  2;
26. algorithm        Newton;
27. integrator       LoadControl0.1;
28. analysis         Static;
29. analyze          10;
```

Lines 22–29 denote the static analysis for applying gravity loading.

```
30. wipeAnalysis;
31. constraints Plain;
32. numberer Plain;
33. system BandGeneral;
34. test NormDispIncr 1.0e-8 100 2;
35. algorithm Newton;
36. integrator Newmark 0.5 0.25;
37. analysis Transient;
38. analyze 3000 $pDeltT
39. wipe;
```

Lines 30–39 perform 3000-step dynamic analysis with 0.001 s per step. As for the two files to define the bridge and train, the following are the Tcl commands for the bridge.

```
1. global NodeID
2. if { [file exists DataB] == 0 } {file mkdirDataB}
```

Line 1 defines a global variable used in main.tcl. Line 2 establishes a folder to store the recorded bridge and track information.

```
3.  set LenE1.5;
4.  set H     0.05;
5.  for     {set i 1}     {$i<=81}     {incri 1}     {
6.  node    [expr $i]     [expr ($i-1)*$LenE]     $H
7.  lappendNodeID     [expr $i]
8.  }
9.  for     {set i 101}     {$i<=181}     {incri 1}     {
10. node    [expr $i]     [expr ($i-101)*$LenE]     0.0;
11. }
12. for     {set i 101}     {$i<=130}     {incri 1}     {
13. fix    [expr $i]    1    1    1
14. }
15. for     {set i 152}     {$i<=181}     {incri 1}     {
16. fix    [expr $i]    1    1    1
17. }
18. fix    131    1    1    0;
19. fix    151    0    1    0;
20. fix 1    1 0 1;
21. fix    81    1    0    1;
```

Lines 3–17 define the nodes of the rail and the bridge. Lines 18–21 denote the boundary condition.

```
22. set Ec  2.943e9;
23. set AB  7.94;
24. set IzB 2.88;
25. section Elastic  2  $Ec  $AB  $IzB;
```

Lines 22–25 define the material and section of the bridge element.

```
26. set AR         77.45e-4;
27. set E_rail  2.06e11;
28. set Iz_rail  [expr 2*2.037e-5];
```

Lines 26–28 define the material of the rail element.

```
29. set Krb [expr 2*6.58e7];
30. set Crb [expr 2*3.21e4];
31. set Arb  $H;
32. set Erb [expr $Krb]
33. uniaxialMaterial Elastic  101  $Erb
34. uniaxialMaterial Viscous  201  $Crb  1
35. geomTransf Linear 2;
```

Lines 29–35 define the parameters of spring damper between track and bridge.

```
36. for {set i 131} {$i<151} {incri 1} {;
37. element  dispBeamColumn  $i  $i  [expr $i+1]  5  2  2
     -mass 1.2e4 -cMass;
38. }
39. for {set i 1} {$i<81} {incri 1} {
40. element elasticBeamColumn  $i  $i [expr ($i+1) ]  $AR
     $E_rail $Iz_rail  2  -mass  [expr 2*51.5]  -cMass
41. }
42. for {set i 1} {$i<=81} {incri 1} {
43. element truss [expr $i+6000] [expr $i] [expr $i+100] $Arb 101
44. element truss [expr $i+7000] [expr $i] [expr $i+100] $Arb 201
45. }
```

Lines 36–45 define the element of the bridge, rail, and the spring dampers between the rail and the bridge.

```
46. recorder Node -file DataB/RailDisp.txt  -time -node 16 41 66
     -precision 16  -dof 2  disp;
47. recorder Node -file DataB/RailAccel.txt  -time -node 16 41 66
     -precision 16  -dof 2  accel;
48. recorder Node -file DataB/BridgeDisp.txt  -time -node  141
     -precision 16  -dof 2  disp;
49. recorder Node -file DataB/BridgeAccel.txt -time -node  141
     -precision 16  -dof 2  accel;
```

Lines 46–49 record the responses of the bridge and the rail. The above is the definition of the bridge model. The vehicle model is defined in train.tcl as follows.

```
1. if { [file exists DataT] == 0 } {file mkdirDataT}
2. set g 9.801;
3. set Mt 5.2e4;
4. set JMt 2.31e6;
5. set Mb 3.2e3;
6. set JMb 3.12e3;
7. set MWheel 1.4e3;
8. set ytranslation [expr 0.05+$pRWheel];
9. set xtranslation [expr 10.25+$pInitLocation];
10. node 2001 [expr -10.25+$xtranslation] [expr
    0+$ytranslation]
11. node 2002 [expr -7.75+$xtranslation] [expr
    0+$ytranslation]
12. node 2003 [expr 7.75+$xtranslation] [expr
    0+$ytranslation]
13. node 2004 [expr 10.25+$xtranslation] [expr
    0+$ytranslation]
14. node 2005 [expr -10.25+$xtranslation] [expr
    0.22+$ytranslation]
15. node 2006 [expr -9+$xtranslation] [expr
    0.22+$ytranslation]
16. node 2007 [expr -7.75+$xtranslation] [expr
    0.22+$ytranslation]
17. node 2008 [expr 7.75+$xtranslation] [expr
    0.22+$ytranslation]
18. node 2009 [expr 9+$xtranslation] [expr
    0.22+$ytranslation]
19. node 2010 [expr 10.25+$xtranslation] [expr
    0.22+$ytranslation]
20. node 2011 [expr -9+$xtranslation] [expr
    0.52+$ytranslation]
21. node 2012 [expr 0+$xtranslation] [expr
    0.52+$ytranslation]
```

```
22. node 2013 [expr 9+$xtranslation] [expr
     0.52+$ytranslation]
```

Lines 10–22 define the nodes of the vehicle.

```
23. mass 2001 $MWheel $MWheel 0.0
24. mass 2002 $MWheel $MWheel 0.0
25. mass 2003 $MWheel $MWheel 0.0
26. mass 2004 $MWheel $MWheel 0.0
27. mass 2006 $Mb $Mb $JMb
28. mass 2009 $Mb $Mb $JMb
29. mass 2012 $Mt $Mt $JMt
```

Lines 23–29 define the mass of the nodes of the vehicle.

```
30. fix 2001 1 0 1
31. fix 2002 1 0 1
32. fix 2003 1 0 1
33. fix 2004 1 0 1
34. fix 2005 1 0 0
35. fix 2006 1 0 0
36. fix 2007 1 0 0
37. fix 2008 1 0 0
38. fix 2009 1 0 0
39. fix 2010 1 0 0
40. fix 2011 1 0 0
41. fix 2012 1 0 0
42. fix 2013 1 0 0
```

Lines 30–42 define the boundary condition of the vehicle model, where the wheel node has only vertical dof, and the bogie and car body nodes have only vertical and rotational dofs.

```
43. equalDOF 2012 2011  3
44. equalDOF 2012 2013  3
45. equalDOF 2006 2005  3
```

```
46. equalDOF 2006 2007  3
47. equalDOF 2009 2008  3
48. equalDOF 2009 2010  3
```

Lines 43–48 are used to construct a multi-point constraint between nodes.

```
49. set E  2.06e11;
50. set po 0.3 ;
51. set Kv1 1.87e6;
52. set Cv1 5.0e5
53. set Kv2 1.72e6
54. set Cv2 1.96e5
55. set Av1 0.2200;
56. set Av2 0.3000;
57. set Ev1 [expr $Kv1]
58. set Ev2 [expr $Kv2]
59. set Izb [expr 4.05e7];
60. set Ab  [expr 0.54]
61. set At  [expr 8.4]
62. set Izt [expr 3.0e7]
63. uniaxialMaterial Elastic  801 $Ev1
64. uniaxialMaterial Elastic  802 $Ev2
65. uniaxialMaterial Viscous 701  $Cv1  1
66. uniaxialMaterial Viscous 702  $Cv2  1
```

Lines 49–66 define the parameters of the bogie and the car body.

```
67. element elasticBeamColumn 2001 2005 2006 $Ab $E $Izb 2
68. element elasticBeamColumn 2002 2006 2007 $Ab $E $Izb 2
69. element elasticBeamColumn 2003 2008 2009 $Ab $E $Izb 2
70. element elasticBeamColumn 2004 2009 2010 $Ab $E $Izb 2
71. element elasticBeamColumn 2005 2011 2012 $At $E $Izt 2
72. element elasticBeamColumn 2006 2012 2013 $At $E $Izt 2
73. element truss 2007  2001  2005  $Av1  801
74. element truss 2008  2002  2007  $Av1  801
75. element truss 2009  2003  2008  $Av1  801
```

```
76. element truss 2010 2004 2010 $Av1 801
77. element truss 2011 2006 2011 $Av2 802
78. element truss 2012 2009 2013 $Av2 802
79. element truss 2013 2001 2005 $Av1 701
80. element truss 2014 2002 2007 $Av1 701
81. element truss 2015 2003 2008 $Av1 701
82. element truss 2016 2004 2010 $Av1 701
83. element truss 2017 2006 2011 $Av2 702
84. element truss 2018 2009 2013 $Av2 702
```

Lines 67–84 define the element of the car body, bogie, and the spring damper between the car body and the bogie.

```
85. pattern Plain 1 Linear {
86. load 2001 0.0 [expr -$MWheel*$g] 0.0;
87. load 2002 0.0 [expr -$MWheel*$g] 0.0;
88. load 2003 0.0 [expr -$MWheel*$g] 0.0;
89. load 2004 0.0 [expr -$MWheel*$g] 0.0;
90. load 2006 0.0 [expr -$Mb*$g] 0.0;
91. load 2009 0.0 [expr -$Mb*$g] 0.0;
92. load 2012 0.0 [expr -$Mt*$g] 0.0;
93. }
```

Lines 85–93 define the gravity loading of car body and bogie.

```
94. recorder Node -file DataT/WnodeDisp.txt -time -node 2001
    2002 2003 2004 -precision 16 -dof 2 disp;
95. recorder Node -file DataT/WnodeAccel.txt -time -node 2001
    2002 2003 2004 -precision 16 -dof 2 accel;
96. recorder Element -file DataT/SpringEleF.txt -time -ele 2007
    2008 2009 2010 -precision 16 localForce;
97. recorder Node -file DataT/bodyDisp.txt -time -node 2011 2012
    2013 -precision 16 -dof 2 disp;
98. recorder Node -file DataT/bodyVel.txt -time -node 2011 2012
    2013 -precision 16 -dof 2 vel;
99. recorder Node -file DataT/bodyAccel.txt -time -node 2011 2012
    2013 -precision 16 -dof 2 accel;
```

```
100. recorder Node -file DataT/bogieDisp.txt -time -node 2005 2006
     2007 2008 2009 2010 -precision 16  -dof  2  disp;
101. recorder Node -file DataT/bogieVel.txt  -time -node 2005 2006
     2007 2008 2009 2010  -precision 16  -dof  2  vel;
102. recorder Node -file DataT/bogieAccel.txt -time -node 2005 2006
     2007 2008 2009 2010 -precision 16   -dof  2  accel;
103. recorder Element -file DataT/bodyForce.txt  -time -ele 2005  2006
     -precision 16 localForce;
104. recorder Element -file DataT/bogieForce.txt  -time -ele 2001  2002
     2003 2004 -precision 16 localForce;
```

Lines 94–104 record the responses of the wheel, car body, and the bogie.

Analysis results: The vertical displacements of vehicle body are shown in Fig. 13.3. It is clear that when the train passes the middle of the bridge,

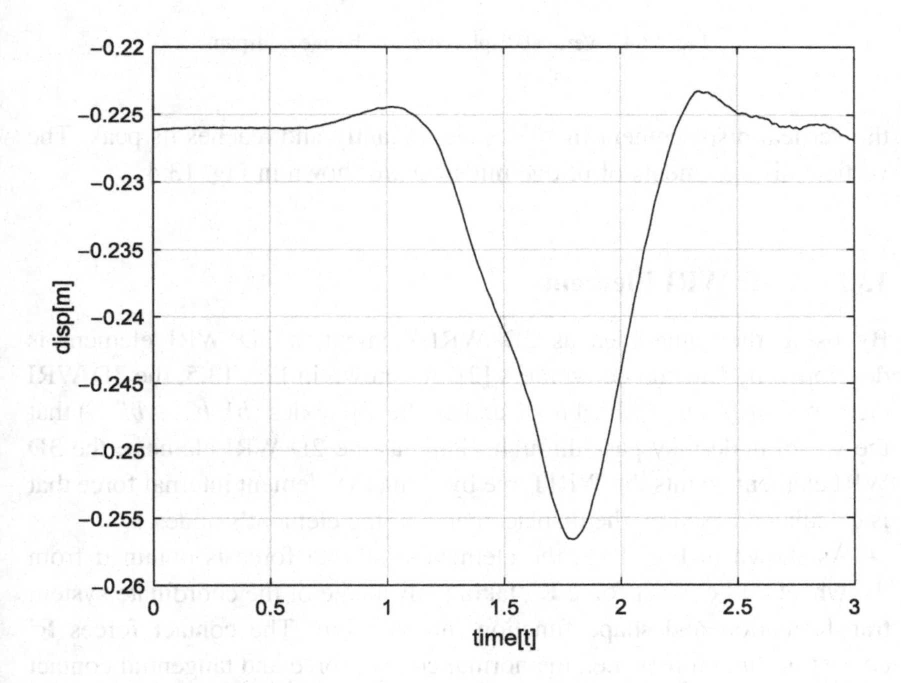

Fig. 13.3 Vertical displacement of vehicle body.

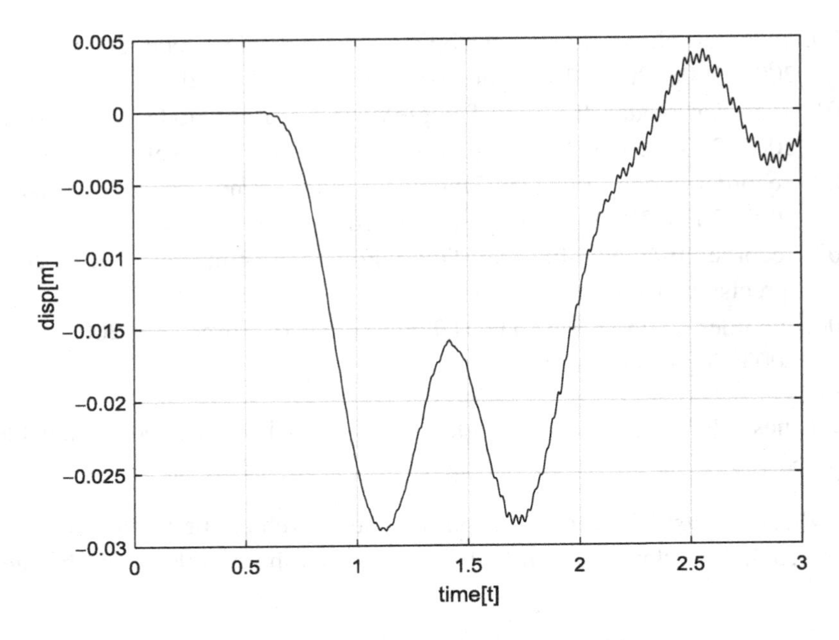

Fig. 13.4 Vertical displacement of bridge midspan.

the vertical displacement increases significantly and reaches its peak. The vertical displacements of bridge midspan are shown in Fig. 13.4.

13.2 A 3D WRI Element

By using the same idea as 2D WRI element, a 3D WRI element is developed by Liu and co-workers [2]. As shown in Fig. 13.5, the 3D WRI element consists of a wheel node and all the rail nodes $(b1, b2 \ldots bi \ldots)$ that the wheel node may pass through. Same as the 2D WRI element, the 3D WRI element counts the WRI force by using the element internal force that is calculated based on the displacements of the element's nodes.

As shown in Fig. 13.6, the element's internal force is obtained from the wheel–rail contact force $\mathbf{F}^c$, taking advantage of the coordinate system transformation and shape function interpolation. The contact forces $\mathbf{F}^c$ consist of three forces, i.e., the normal contact force and tangential contact forces along x-axis and y-axis directions.

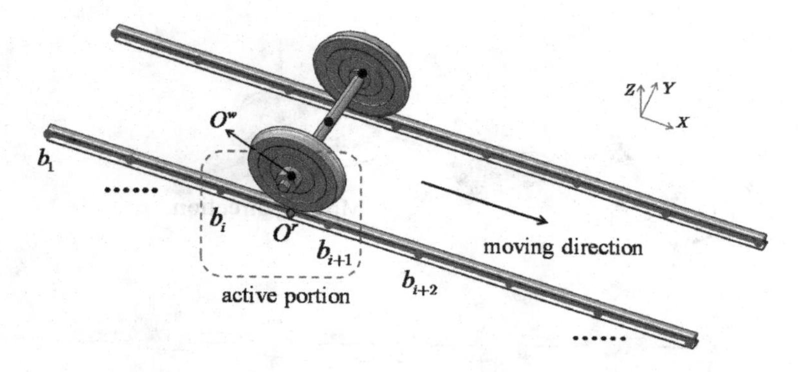

Fig. 13.5 Model of the 3D WRI element.

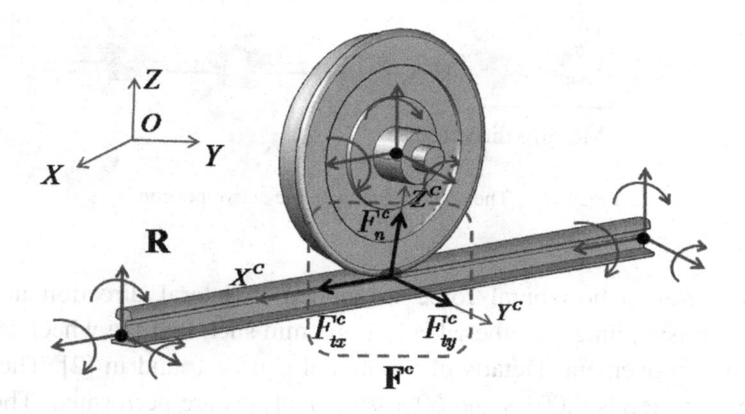

Fig. 13.6 The contact forces in the active portion.

The normal contact force is obtained by the nonlinear Hertz theory and the tangential contact forces are obtained by Polach formulation. The details of the formula can be referred in [2]. It is worth mentioning that the source code of the 3D WRI element has not been uploaded to the official website of OpenSees, however, the executive file and tcl examples can be found at https://github.com/OpenSeesXMU.

The following is an example of a wheelset moving over a straight rigid track.

As shown in Fig. 13.7, a single wheelset is running on a 120 m rigid straight rail at a constant speed of 20 m/s. The wheel and rail profiles used are the ML95 and UIC50. Before the dynamic analysis of

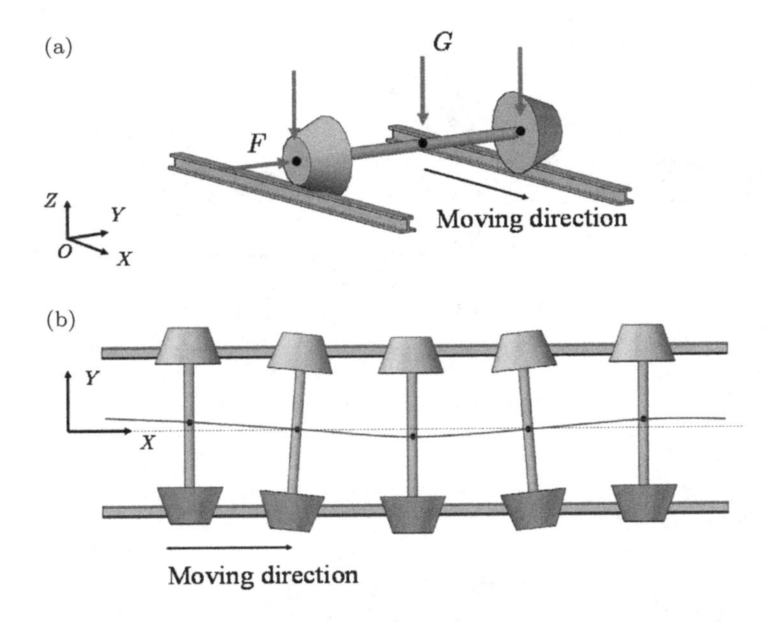

Fig. 13.7 The contact forces in the active portion.

the wheelset, a horizontal force is applied in lateral direction and an original misalignment of the wheelset is 2 mm such that the wheel set has a hunting movement. Details of the model can be found in [2]. The size of the time step is 0.001 s and 6000-step analyses are performed. The Tcl commands are as follows:

```
1.  global R0 la lbEwmiuWheel;
2.  set R0 0.43;
3.  set zFrame 0.46;
4.  set zCar 1.849;
5.  set la 2.1;
6.  set lb 11.1;
7.  set Ew 2.06e11;
8.  set miuWheel 0.296;
9.  set Mw 933;    set Iwx 461.4;   set Iwy 61.6;   set Iwz 461.4;
```

```
10. set Ma 176;    set Iax 144.5;   set Iay 2.2;    set Iaz 144.5;
11. set Mt 1982;   set Itx 1398.5; set Ity 2667;    set Itz 2667;
12. set Mc 11160; set Icx 14952.9; set Icy 225365.4; set Icz 224994.9;
13. set Mwheel  407.629373;
14. set Mcenter 117.741254;
15. set y0 0.7523;#7523
16. set y1 0.90;
```

Lines 1–16 define the basic information of the wheelset.

```
17. node 1      0  -$y0  $R0;
18. node 1001 0    $y0  $R0;
19. node 51     0   -$y1  $R0;
20. node 1051 0    $y1  $R0;
21. node 2001 0  0   $R0 -mass $Mw $Mw $Mw $Iwx $Iwy $Iwz;
22. fix 2001 0 0 0 0 1 0;
```

Lines 17–22 define the node of the wheelset and the boundary condition of the wheelset center.

```
23. SET ARigid 1.0e3; set GRigid 7.9E10; set ERigid 2.0e11;
24. set IxRigid 0.1;  set IyRigid 0.1;   set IzRigid 0.1;
25. set transf1 1;
26. geomTransf Linear $transf1 0 0 1;
```

Lines 23–26 define the parameters of the wheelset axis and the coordinate system.

```
27. element elasticBeamColumn 2001     1      2001 $ARigid $ERigid
    $GRigid $IxRigid $IyRigid $IzRigid $transf1;
28. element elasticBeamColumn 2101     2001   1001 $ARigid
    $ERigid $GRigid $IxRigid $IyRigid $IzRigid $transf1;
29. element elasticBeamColumn 2051     1      51   $ARigid $ERigid
    $GRigid $IxRigid $IyRigid $IzRigid $transf1;
30. element elasticBeamColumn 2151     1001   1051 $ARigid
    $ERigid $GRigid $IxRigid $IyRigid $IzRigid $transf1;
```

Lines 27–30 define the element of the wheelset.

```
31. pattern Plain 2 { Series -time { 0.0 0.01 10000.0 }
32. -values { 0.0 1.0 1.0 } } {
33. set g 9.801;
34. load 2001 0 0 [expr -$Mw*$g] 0 0 0;
35. }
```

Lines 31–35 define the gravity loading of the wheelset.

```
36. pattern Plain 3 { Series -time { 0.0 0.01 0.02
    0.200001 800.0 }
37. -values { 0.0 0 2.48e-4 0.0 0 } } {
38. load 2001 0 [expr $Mw*$g] 0 0 0 0;
39. }
```

Lines 36–39 define a horizontal force applied in lateral direction to make an original misalignment of the wheelset of 2 mm.

```
40. recorder Node -file ex1/W1DispL.txt -time -node 1
    -precision 16 -dof 1 2 3 4 6 disp;
41. recorder Node -file ex1/W1VelL.txt -time -node 1
    -precision 16 -dof 1 2 3 4 6 vel;
42. recorder Node -file ex1/W1AccelL.txt -time -node 1
    -precision 16 -dof 1 2 3 4 6 accel;
```

Then is the rail.tcl.

```
1. global NodeIDLeftNodeIDRightDataBAr Gr IryEr Jr
   IrzmrailLele g wbmiuRailzrNnodetransfRail;
2. set Nnode 200;
3. set zr 0.15;
4. set wb 1.505;
5. set Lele 0.6;
6. for {set i 1} {$i<=$Nnode} {incri 1} {
7. node [expr $i+10000] [expr ($i-1)*$Lele] [expr-$wb/2] [expr
   -$zr]
8. lappendNodeIDLeft [expr $i+10000]
9. node [expr $i+50000] [expr ($i-1)*$Lele] [expr $wb/2] [expr
   -$zr]
```

```
10. lappendNodeIDRight   [expr $i+50000]
11. }
```

Lines 6–11 define the rail nodes sequence, where the "lappend" in lines 8 and 10 is used to generate an array. Take line 8 as an example, NodeIDLeft defines the name of the array, [expr $i+10000] defines the values in the array.

```
12. for {set i 1} {$i<=$Nnode} {incri 1} {
13. for {set i 1} {$i<=$Nnode} {incri 1} {
14. fix [expr $i+10000] 1 1 1      1 1 1;
15. fix [expr $i+50000] 1 1 1      1 1 1;
16. }
17. set transfRail 2;
18. geomTransf Linear $transfRail 0 0 1;
```

Lines 12–18 define the boundary condition of the rail nodes and the coordinate system.

```
19. set Er 2.06e11;
20. set miuRail 0.296;
21. set Ar 7.745e-3;
22. set Gr [expr $Er/(1+$miuRail)/2.0];
23. set Jr 2.104e-6;
24. set Irz 5.24e-6;
25. set Iry 3.217e-5;
26. set mrail 60.64;
27. for {set i 1} {$i<$Nnode} {incri 1} {
28. element elasticBeamColumn [expr $i+10000] [expr $i+10000]
    [expr $i+10001] $Ar $Er $Gr $Jr  $Iry $Irz $transfRail -mass
    $mrail
29. element elasticBeamColumn [expr $i+50000] [expr $i+50000]
    [expr $i+50001] $Ar $Er $Gr $Jr $Iry $Irz $transfRail -mass
    $mrail
30. }
```

Lines 19–30 define the element of the rail.

Finally, the main.tcl is as follows:

```
1.  wipe ;
2.  if { [file exists ex1] == 0 } {
3.  file mkdir ex1;
4.  }
5.  model basic -ndm 3 -ndf 6;
6.  source rail.tcl;
7.  source wheelset.tcl;
```

Lines 6–7 include the two model files above.

```
8.  set dT 0.001;
9.  set pVel 20;
10. set nL 20;
11. set omiga [expr $pVel/$R0*1.0];
12. set pInitLocation 0.0;
13. set Loc1 [expr 0+$pInitLocation]
14. set Loc2 [expr $la+$pInitLocation]
15. set Loc3 [expr $lb+$pInitLocation]
16. set Loc4 [expr $la+$lb+$pInitLocation]
17. set WheelP "0 0 0 0 0 0.25 $omiga $pVel $Ew  $miuWheel
    $R0";
18. set RailL  "0 0 $zr -0.05 0 0 $Er $miuRail $Iry $Irz $Jr $Ar";
19. set RailR  "0 0 $zr  0.05 0 0 $Er $miuRail $Iry $Irz $Jr $Ar";
20. set DelList "0 0 0";
21. set DelLocList "14.5 15 15.5";
22. set NodeListL1 "[concat 1 $NodeIDLeft]";
23. set NodeListR1 "[concat1001 $NodeIDRight]";
```

Lines 8–23 define the basic parameters of the model. It is worth mentioning that lines 22 and 23 form a new array of wheel node and rail nodes.

```
24. element WheelRail 200001 $dT $Loc1 $nL -WPara $WheelP -
    RPara $RailR -NodeList $NodeListR1 UIC50Fine.txt irregRht.txt
    ML95_FineR.txt R;
```

25. element WheelRail 100001 $dT $Loc1 $nL -WPara $WheelP -RPara $RailL -NodeList $NodeListL1 UIC50Fine.txt irregLft.txt ML95_FineL.txt L;

Lines 24–25 define the 3D WRI element. Take line 24 as an example, 200001 denotes the tag of the element. dT denotes the size of the time step. Loc1 denotes the initial location of the wheel node. nL denotes the numbers of the gravity loading steps. WheelP can refer to line 17. The first five items represent the coordinate of the wheel node in the wheel coordinate system. 0.25 represents the friction coefficient. omiga, pVel, Ew, miuWheel, and R0 represent wheel rotation angular speed, wheel travel speed, wheel elastic modulus, wheel Poisson's ratio, and wheel radius, respectively. NodeListR1 can refer to line 19. Er, miuRail, Iry, Irz, Jr, and Ar represent the rail elastic modulus, rail Poisson's ratio, cross-sectional moment of inertia in the y direction, cross-sectional moment of inertia in the *z* direction, rail moment of inertia, and rail cross-sectional area. UIC50Fine.txt, irregLft.txt, and ML95_FineL.txt L denote the profile point coordinates of the wheel, track irregularity, and the profile point coordinates of the rail.

26. recorder Element -file ex1/eleForceR1.txt -time -ele 200001 -precision 10 localForce;
27. recorder Element -file ex1/eleForceL1.txt -time -ele 100001 -precision 10 localForce;
28. set error 1.0e-3;
29. set maxIt 20;
30. constraints Transformation;
31. numberer Plain;
32. system BandGeneral;
33. test NormUnbalance $error $maxIt 1;
34. algorithm Newton;
35. integrator LoadControl 0.001;
36. analysis Static;
37. set startT [clock seconds];
38. analyze $nL;
39. Lines 28−38 define the gravity loading.
40. wipeAnalysis;

```
41. constraints Transformation;
42. numberer Plain;
43. system BandGeneral;
44. test NormUnbalance $error $maxIt 1;
45. algorithm Newton;
46. integrator Newmark 0.5 0.25;
47. analysis Transient;
48. analyze 6000 $dT;
49. set endT [clock seconds]
50. puts "Execution time: [expr $endT-$startT] seconds."
51. wipe;
```

Lines 39–50 define the dynamic analysis.

Analysis results: The lateral displacement and the lateral force of the wheel node (wheel node 1, line 17 in wheelset.tcl) are shown in Figs. 13.8 and 13.9, respectively. It is clear that the amplitude of lateral displacement increases gradually and then remains "regular" when the collision between the flange and the rail occurs regularly. Whenever the collision occurs, the lateral force jumps abruptly.

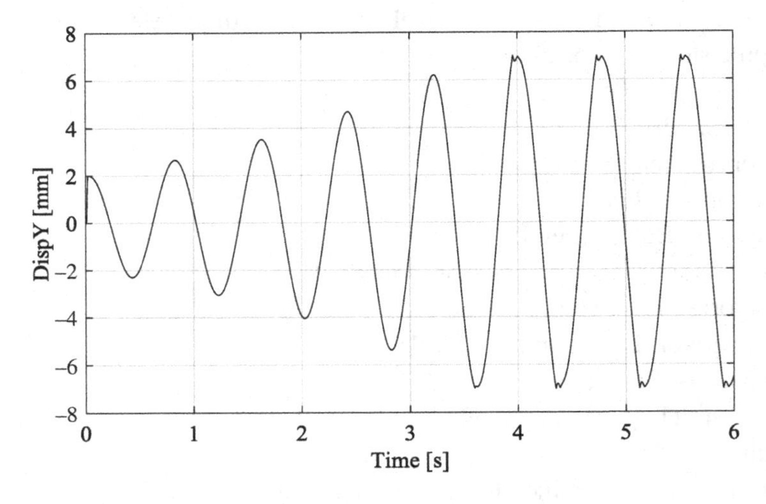

Fig. 13.8 Lateral displacement of wheel node.

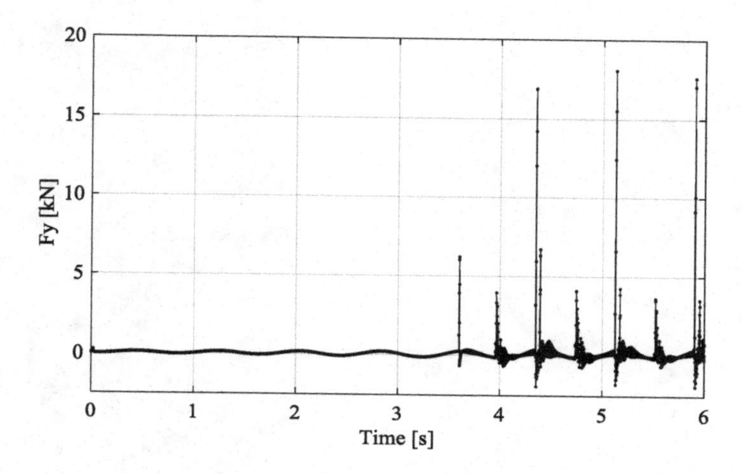

Fig. 13.9 Lateral force of wheel node.

References

[1] Gu, Q., Liu, Y., Guo, W., Li, W., Yu, Z., *et al.* (2019). A practical wheel-rail interaction element for modeling vehicle-track-bridge systems. *Int. J. Struct. Stab. Dyn. 19*(2), 1950011. DOI 10.1142/S0219455419500111.

[2] Liu, Y., Montenegro, P., Gu, Q., Guo, W., Calçadab, R. João Pombo. (2021). A practical three-dimensional wheel-rail interaction element for dynamic response analysis of vehicle-track systems. *Computers & Structures, 254,* 106581. DOI 10.1016/j.compstruc.2021.106581.

Chapter 14

Reliability Analysis

14.1 RORM, SORM, and Sampling Analyses

The theory about reliability analysis can be found in Refs. [1] and [2], and this chapter presents the application examples of reliability analysis.

(1) First-order reliability analysis

This example is about the first-order reliability analysis, i.e., FORM analysis. The performance function is defined explicitly in terms of three random variables, i.e.,

$$g(x) = par_1 - \frac{x_2}{100x_3} - \left(\frac{x_1}{20x_3}\right)^2 \tag{14.1}$$

where $par_1 = 0.8$ is a deterministic parameter. The $x_1 - x_3$ are three random variables with lognormal and uniform probability distributions. The correlation between the random variables is specified by a correlation matrix:

$$[\rho_{ij}] = \begin{bmatrix} 1.0 & 0.4 & 0.3 \\ 0.4 & 1.0 & 0.2 \\ 0.3 & 0.2 & 1.0 \end{bmatrix} \tag{14.2}$$

The Tcl commands are as follows:

```
1. randomVariable 1 lognormal  500.0    100.0
2. randomVariable 2 lognormal  2000.0   400.0
3. randomVariable 3 uniform    5.0      0.5
```

Lines 1–3 define the three random variables. The first random variable x_1 is assigned a lognormal probability distribution with a mean of 500 and a standard deviation of 100. The second random variable x_2 is similarly defined. The last random variable x_3 is assigned the uniform probability distribution with a mean of 5 and a standard deviation of 0.5.

```
4. correlate 1 2 0.4
5. correlate 1 3 0.3
6. correlate 2 3 0.2
```

Lines 4–6 describe the correlation between the random variables. For example, line 4 defines a correlation coefficient of 0.3 between random variables with tags 1 and 2.

```
7. set a "{x_2}/(100.0*{x_3})"
8. set b "{x_1}/(20.0*{x_3})"
9. performanceFunction 1 "1.0 - $a - $b*$b"
```

Lines 7–8 define two variables: $a = \frac{x_2}{100x_3}$ and $b = \frac{x_1}{20x_3}$, see Ref. [1]. Line 9 defines the performance function in Eq. (14.2).

```
10. probabilityTransformation Nataf -print 0
11. reliabilityConvergenceCheck Standard -e1 1.0e-3 -e2 1.0e-3 -
    print 1
12. functionEvaluator Tcl
13. gradGEvaluator FiniteDifference -pert 1000
```

Line 10 specifies how to transform random variables between the original and standard normal spaces. Currently, only Nataf transformation is available [2]. Line 11 defines how to check whether the reliability analysis has converged by setting a convergence criteria and target tolerance. Two convergence criteria must be met. The first convergence check ***e1*** determines the closeness of the design point (DP) to the limit-state surface. The second convergence check ***e2*** determines how closely the gradient vector points toward the origin in the standard normal space. Line 12 specifies how to evaluate (i.e., compute the value of) the performance functions (also denoted as limit state functions LSFs or g-functions) for a given realization of the random variables. "***Tcl***" means that only random variables and variables in TCL can be included in the performance function. Other evaluators are available when FE software is used to calculate the performance functions, as will be shown later. Line 13 uses forward finite difference method to calculate the gradients of the performance functions and the optional ⟨**-pert *$arg1***⟩ specifies the perturbation size as 1/1000, with the default perturbation size being 1/1000.

14. rootFinding Secant -maxIter 50 -tol 1.0e-2 -maxStepLength 1.0

Line 14 uses the secant method for the root-finding algorithm to find the optimum point, i.e., the DP, by taking advantage of the previously defined objects of gradient and limit-state function. The maximum iteration number is ***50*** and the target tolerance is ***1.0e-2*** (which is calculated by dividing the current value of the performance function by the initial value of the performance function). The argument ***1.0*** is used to constrain the length of the steps.

15. stepSizeRule Fixed -stepSize 1.0

In this line, a **stepSizeRule** command calculates the step size along a search direction in the line search algorithm. Line 15 uses the ***Fixed*** selection and specifies a constant step size 1.0 (given by users).

16. searchDirectionGradientProjection

In this line, a ***searchdirection*** object specifies the algorithm to get the search direction for searching the design point (DP). The GradientProjection is the algorithm for search direction [1].

17. startPoint Mean
18. findDesignPoint StepSearch -maxNumIter 100

Line 17 denotes the initial value is the mean of the random variables for the DP search algorithm. Line 18 uses a step-by-step search scheme to search for the DP and the maximum number of iterations for DP search is 100.

19. runFORMAnalysisFORMoutput.out

Line 19 specifies the first-order reliability analysis being performed, and the analysis results are stored in the output file "FORMoutput.out".

(2) Second-order reliability analysis

In this section, a second-order reliability analysis is performed and the corresponding Tcl commands are the same as FORM analysis in this section except the following two lines are performed after line 19,

```
20 findCurvatures firstPrincipal --exe
21 runSORMAnalysis SORMoutput.out
```

Line 20 performances a **findCurvatures** command to record the principal curvature [3], e.g., to get the first principal curvature. Line 21 is used to perform the second-order reliability (SORM) analysis and the results are stored in the output file "SORMoutput.out". It is worth noting that the current SORM analysis in line 21 requires an FORM analysis (line 20) and a findCurvatures (line 21) to be performed before the SORM analysis.

(3) Importance sampling analysis

An importance sampling analysis around the design point can be performed and the corresponding Tcl commands from lines 1 to 17 are the same as FORM analysis in this section. The following four lines are added after line 17 (i.e., previous lines 18 and 19 are replaced by the new ones).

```
18. findDesignPoint StepSearch -maxNumIter 100 -
    printDesignPointX designPoint1.txt
19. runFORMAnalysis FORMoutput.out
20. startPoint -file designPoint1.txt
21. runImportanceSamplingAnalysis SIMULATIONoutput1.out -
    type failureProbability -variance 1.0 -maxNum 10000 - targetCOV
    0.01 -print 0
```

Line 18 specifies the method to search for the DP by a step-by-step search scheme, and the design point is stored in the output file "designPoint1.txt" (this is necessary for importance sampling analysis). Line 19 specifies the first-order reliability analysis is performed, and the analysis results are stored in the output file "FORMoutput.out". Line 20 specifies the initial value used for the DP search algorithm, which is a specified initial DP vector or data in a file. Line 21 performs an importance sampling (IS) method. The analysis results will be recorded in the output file named "SIMULATIONoutput1.out". Where ***–variance*** denotes the standard deviation of the sampling distribution is ***$var*** (default = 1.0), ***-maxNum*** denotes the maximum number of simulations is 10000, ***-targetCOV*** denotes the target coefficient of variation of the estimate is 0.01 (default = 0.05), ***-print*** is a print flag with the following meaning: 0, the status of the sampling analysis is not printed to the screen or file; 1, the status after each sample is printed only to the screen; 2, the status after each sample is printed only to the screen, while necessary information is printed into a file as well such that the sampling analysis may be restarted (i.e., the analysis will continue from the current point at the next run). Selection 2 is very useful for computational cases that need a large number of FE simulations.

14.2 Introduction to an Open Source Reliability Analysis Software

Recently, the existing reliability framework in OpenSees has been enhanced by the authors, Professors Sanjay Govindjee, Ziqi Wang, and other co-workers, based on which a stand-alone open-source reliability analysis software (ORS) has been developed for structural reliability analysis. This chapter will introduce the application and usage of the software. The source code, examples, and user's guide of the ORS can be found in the link: https://github.com/OpenSeesXMU.

The ORS can perform time-invariant or time-variant reliability analyses, e.g., first-order reliability analysis (FORM), second-order reliability analysis (SORM), Monte Carlo sampling (MCS), importance sampling (IS), subset simulation analysis and mean upcrossing rate analysis. The ORS can be used to perform reliability analysis for complicated engineering problems by combining with other finite element analysis software (e.g., OpenSees or Abaqus).

A number of numerical examples are presented in this chapter, starting from simple problems and proceeding to relatively comprehensive examples involving structural reliability analysis. The analysis types and corresponding execution commands are summarized in Table 14.1.

Table 14.1 Analysis types and corresponding execution commands of reliability analysis in ORS.

Tag	Analysis types	Executing commands
1	FORM Analysis	runFORMAnalysis
2	SORM Analysis	runSORMAnalysis
3	Importance Sampling Analysis	runImportanceSamplingAnalysis
4	Orthogonal Plane Sampling Analysis	runOrthogonalPlaneSamplingAnalysis
5	Monte Carlo Sampling	runMonteCarloResponseAnalysis
6	Subset Simulation Analysis	runSubsetSimulationAnalysis
7	Mean Upcrossing Rate Analysis	runOutCrossingAnalysis

14.3 Time-Invariant Reliability Analysis

(1) Orthogonal plane sampling analysis

An orthogonal plane sampling analysis is performed for a single degree of freedom system. The truss element with steel01 material is used to model the system. Tcl commands of the reliability analysis are as follows:

```
1   randomVariable 1 lognormal 210000.0 [expr 0.05*210000.0]
    210000.0
2   randomVariable 2 lognormal 355.0 [expr 0.05*355.0]  355.0
3   randomVariable 3 lognormal 337500.0  [expr 0.05*337500.0]
    337500.0
4   performanceFunction 1 "1.85-{u_2_1}"
```

Line 4 defines the performance function where $\{u\_2\_1\}$ is horizontal response of node 2. The response will be calculated by using other finite element analysis software (e.g., OpenSees).

```
5.  probabilityTransformationNataf    -print 0
6.  randomNumberGeneratorCstdLib
```

Line 6 employs the standard library function in C++ to generate the random number.

```
7.  reliabilityConvergenceCheck  Standard  -e1 1.0e-3 -e2 1.0e-3 -
    print 1
8.  gFunEvaluatorOpenSeesTimeinvariant -file test9.tcl
```

Line 8 specifies how to evaluate (i.e., compute the value of) the performance functions for a given realization of the random variables. When ***OpenSees*** is specified as g-function evaluator, the OpenSees.exe will be called each time when the performance function needs to be evaluated. In this process, the values of the random variable are created by ORS, recorded in a file named RV.txt, and assigned to the parameters of FE analysis. The response quantities from FE analysis may be included in the performance function. ***Timeinvariant*** denotes that the reliability analysis is time-invariant. The user should provide a file with a name test9.tcl including detailed TCL commands of FE analysis.

9. gradGEvaluatorFiniteDifference -pert 1000
10. searchDirectioniHLRF
11. meritFunctionCheckAdkZhang -multi 2.0 -add 10.0 -factor 0.5

Line 11 defines a merit function used to determine the suitability of a step size. The ***AdkZhang*** selection uses the arguments ***$arg1*** and ***$arg2*** to compute the factor c so that $c > \frac{\|u\|}{\|\nabla G\|}$ by the following equation $c = (arg2)\frac{\|u\|}{\|\nabla G\|} + (arg1)$. Where u is a vector of standard normal variates, and ∇G denotes the gradient operator with respect to u. Default values are ***arg2***=2 and ***arg1***=10. The argument **-factor** is used in the equation f_1 (old and new trial point) $\leq$ \$factor $\lambda f_2(\cdots)$, which is a typical format of a merit function check [1].

12. stepSizeRule Armijo -maxNum 50 -base 0.5 -initial 1.0 2 -print 1 -sphere 1000.0 1.0 1.0

Line 12 calculates the step size along a search direction in the line search algorithm. Where ***50*** represents the maximum number of step size reductions before the errors are accepted; ***0.5*** represents the base number b in the step size value b^k, where $k \geq 0$ is the smallest integer satisfying the merit function check. The optional ***-initial*** and ***-sphere*** flags are available to avoid trial steps too far out in the failure domain. With the ***-initial*** option the user specifies the value b0 = 1.0 after ***numSteps*** trial steps. The alternative ***–sphere*** option allows the user to define a hyper-***sphere***, within which the trial steps are restricted to stay. The sphere is defined in the standard normal space and its radius is ***1000***. The evolution of the

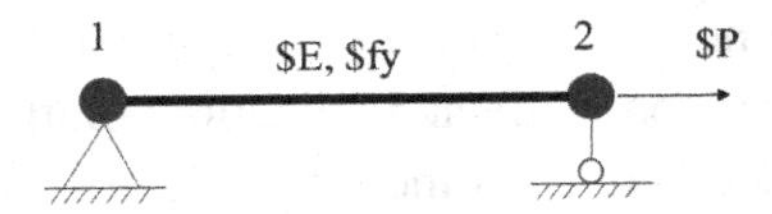

Fig. 14.1 Model of the single degree of freedom system.

radius is based on the input arguments ***1.0*** and ***1.0*** if the design point is not located inside the hyper-sphere.

```
13. startPoint    Mean
14. findDesignPointStepSearch  -maxNumIter 100  -
     printDesignPointX designPoint1.txt
15. runFORMAnalysisFORMoutput.out
16. startPoint  -file designPoint1.txt
17. runOrthogonalPlaneSamplingAnalysis                -fileName
     Orthogonalplaneoutputoutput.out     -maxNum 10000     -type
     failureProbability -targetCOV 0.01 -print 1
```

Line 17 performs an Orthogonal Plane Sampling Analysis. The definition ***-fileName***, ***-maxNum, -type, -targetCOV***, ***-print*** are the same as those in the 'runImportanceSamplingAnalysis' command.

A truss element with steel01 material is used to model the single degree of freedom system, as shown in Fig. 14.1. Tcl commands of the structural response analysis in OpenSees are as follows (named test9.tcl), and detail described of TCL commands for the structural response analysis can be found in [3].

```
1  model basic -ndm 2 -ndf 2
2  set GMScale_sample [open RV.txt "r"];
3  set GMScale_data [read $GMScale_sample];
4  close $GMScale_sample;
5  set E [lindex $GMScale_data 0];
6  set fy [lindex $GMScale_data 1];
7  set P [lindex $GMScale_data 2];
8  node 1 0.0     0.0
9  node 2 5000.0 0.0
10 uniaxialMaterial Steel01  1  $fy  $E  0.02
```

```
11 element truss 1  1 2 1000.0   1
12 fix 1 1 1
13 fix 2 0 1
14 pattern Plain 1 Linear {
15 load 2 $P 0.0
16 }
17 recorder Node nodeDisp.out -time -node 2 -dof 1 disp
18 constraints Plain
19 numberer RCM
20 test NormUnbalance 1.0e-6 25 0
21 integrator LoadControl 0.025 1 0.025 0.025
22 algorithm Newton
23 system ProfileSPD
24 integrator LoadControl  0.02   3   0.02    0.02
25 analysis Static
26 analyze 10
27 set u21 [nodeDisp 2 1]
28 set fo [open PerformanceU.txt w] ;
29 foreachdifferent_content {u21} {
30 puts $fo [set $different_content]
31 }
32 close $fo
```

(2) Monte Carlo sampling

A Monte Carlo sampling analysis is performed for a single degree of freedom system modeled by an elastic beam column element. The material and geometric parameters of the elastic element are determined. The elastic modulus is 2.7×10^{10} Pa and the size of the cross-section is 0.75 m × 0.75 m. A horizontal random load is applied to the top node.

Tcl commands of the reliability analysis are as follows:

```
1   randomVariable 1 normal  5000.0  [expr 0.05*5000.0]  5000.0
2   performanceFunction 1 "0.13-{u_2_1}"
3   probabilityTransformation Nataf   -print 0
4   randomNumberGenerator CStdLib
5   gFunEvaluator OpenSees Timeinvariant   -file test12.tcl
```

Line 5 specifies the performance function evaluated by ***OpenSees***, and the user should provide a file with name test12.tcl with detailed TCL commands of FE analysis.

6. runMonteCarloResponseAnalysis -outPutFilem.out -maxNum 3000 -print 1

Line 6 performs a Monte Carlo analysis. The definitions of ***-outPutFile***, ***-maxNum***, ***-print*** are the same as those in the 'runImportanceSamplingAnalysis' command.

Tcl commands of the structural response analysis in OpenSees are as follows (named test12.tcl):

```
1  model basic -ndm 2 -ndf 3
2  node 1 0.0 0.0
3  node 2 0.0 6000.0
4  fix 1 1 1 1
5  fix 2 0 1 1
6  geomTransf Linear 1
7  element elasticBeamColumn 1 1 2 562500 3.0E+04
   2.636719E+010 1
8  set F_channel [open RV.txt "r"];
9  set FData [read $F_channel];
10 close $F_channel;
11 set F [lindex $FData 0];
12 pattern Plain 1 Linear {load 2 $F 0.0 0.0}
13 recorder Node -file DisplacementUnderDeadLoad.txt-time -node
   2 -dof 1 disp
14 constraints Lagrange
15 numberer Plain
16 system BandGeneral
17 test EnergyIncr 1.0e-006 200
18 algorithm Newton
19 integrator LoadControl 0.5
20 analysis Static
21 analyze 2
22 set u21 [nodeDisp 2 1]
```

```
23 set fo [open PerformanceU.txt w] ;
24 foreachdifferent_content {u21} {
25 puts $fo[set $different_content]
26 }
27 close $fo
```

(3) Subset simulation analysis

A subset simulation analysis is performed for a single degree of freedom system [4]. Tcl commands of the reliability analysis are as follows:

```
1 randomVariable 1 normal 5000.0 [expr 0.05*5000.0]
2 performanceFunction 1 "0.13-{u_2_1}"
3 probabilityTransformation Nataf -print 0
4 randomNumberGenerator CStdLib
5 gFunEvaluator OpenSees Timeinvariant -file test12.tcl
6 runSubsetSimulationAnalysis -outPutFilem.out -SamplingType
  HMC -NumSeedSamples 1000 -print 1
```

Line 6 performs a subset simulation analysis. The definitions of ***–outPutFile***, ***-print*** are the same as those in the 'runImportanceSamplingAnalysis' command. ***–SamplingType*** denotes the sampling method in subset simulation process is "HMC", corresponding to the Hamiltonian Monte Carlo method [4]. The optional ***-NumSeedSamples*** denotes the number of samples in each subset is ***1000*** (default=1000).

Tcl commands of the structural response analysis in OpenSees are described as in Section 14.3 (i.e., test12.tcl). The failure probability and the coefficient of variation are Pf = 0.00169, c.o.v = 0.129, respectively.

Next, another analysis method of subset simulation is introduced. The corresponding Tcl commands are the same as the previous case (i.e., "HMC") except Line 6 is modified by the following command:

```
6 runSubsetSimulationAnalysis -outPutFilem.out -SamplingType
  MH 2 -NumSeedSamples 5000 -print 1
```

The sampling method in the subset simulation process is "MH" corresponding to Metropolis Hasting [4]. ***$arg2*** is the special input argument when the MH method is chosen. The value of the ***$arg2*** with the corresponding meaning is as follows: 1. the proposal distribution in the MH sampling method is a standard normal distribution; 2. the proposal distribution is a uniform distribution with mean zero and width two. The optional ***-NumSeedSamples*** denotes the number of samples in each subset is ***5000*** (default=1000).

14.4 Time-Variant Reliability Analysis

(1) Mean upcrossing rate

In time-variant reliability analysis, the FORM approximation of the mean upcrossing rate is computed, and the corresponding Tcl commands of reliability analysis are as follows:

```
1  set totalTime    3.0
2  set numPulses     10
3  set numTimeSteps [expr $numPulses*10]
4  set pi 3.14159265358979
5  set phi0 0.25
6  set deltat [expr $totalTime/$numPulses]
7  for { set i 1 } { $i<= [expr $numPulses] } { incri } {
   randomVariable        $i normal                    0.0       [expr
   pow(2*$pi*$phi0/$deltat,0.5)]0.0
   }
8  performanceFunction 1 "0.0048 - {u_2_1}"
9  probabilityTransformationNataf                -print 0
10 reliabilityConvergenceCheck  Standard      -e1 1.0e-6    -e2  1.0e-
   6  -print 1
11 gFunEvaluator   OpenSees      Timevariant  -analyze
   $numTimeSteps  [expr $totalTime/$numTimeSteps] -file
   reliability_3_aux.tcl
```

Line 11 specifies how to evaluate (i.e., compute the value of) the performance functions for a given realization of the random variables. The OpenSees.exe is called each time when the performance function needs to be evaluated. ***Timevariant*** denotes that the reliability analysis is time-variant. For time-variant reliability analysis, the user may specify the number of steps (i.e., $numTimeSteps) and the analysis time of each step (i.e., $totalTime/$numTimeSteps). The user should also provide a file with name "reliability_3_aux.tcl" for detailed TCL commands of FE analysis.

```
12  gradGEvaluator FiniteDifference   -pert 1000
13  searchDirection iHLRF
14  meritFunctionCheck AdkZhang -multi 2.0 -add 10.0 -factor 0.5
15  stepSizeRule Armijo -maxNum 50 -base0.5  -initial 1.0 2 -print
    0
16  startPoint     Given
17  findDesignPoint StepSearch        -maxNumIter  100 -
    printDesignPointXdesignPointX.out
18  runOutCrossingAnalysis    OutCross.out    -results
    $stepsToStart $stepsToEnd          $freq         [expr
    $numTimeSteps/$numPulses]  -  littleDt 1.0e-6 -Koo;
```

Line 18 is created to perform dynamic reliability analysis by the mean upcrossing rate and the corresponding results are recorded in the output file named ***OutCross.out. $stepsToStart*** and ***$stepsToEnd*** specify the start and end time steps at which the mean upcrossing rate is performed. ***$freq*** determines at which time point the mean upcrossing rate is evaluated between the start and end time steps. These time points are: ***$stepsToStart***, ***$stepsToStart+$freq***, ***$stepsToStart+2*$freq***, ... until ***$stepsToEnd. $sampleFreq*** is the time interval Δt between two neighboring impulses. The ***$little_dt*** ($\delta t$) is the small time increment by which the second design point excitation $f_2(t)$ is obtained through shifting the first design point $f_1(t)$, i.e., $f_2(t) = f_1(t + \delta t)$. ***$Type*** can be specified as either -Koo or -twoSearches. The latter option implies that two design point searches will be performed for each evaluation point, while the former employs the method developed. It is assumed that the user has specified a discretized random process as time series for at least one load pattern and that the corresponding random variable positioners have been created to map random variables into this time series object.

Tcl commands of the structural response analysis (named reliability_3_aux.tcl) in OpenSees are as follows:

```
1.  model basic -ndm 2 -ndf 2
2.  set zzzzero 0;
```

```
3.  set GMScale_sample [open RV.txt "r"];
4.  set GMScale_data [read $GMScale_sample];
5.  close $GMScale_sample;
6.  set fo1 [open RV2.txt w] ;
7.  foreachdifferent_content {"zzzzero"
    "GMScale_data"} {
8.  puts $fo1 [set $different_content]
9.  }
10. close $fo1
11. set M 28.8e3
12. node 1 0.0 0.0
13. node 2 1.0 0.0 -mass $M $M
14. set K 40560.0e3
15. set pi 3.14159265358979
16. uniaxialMaterialHardening 1  $K  734.0e3   0.0 2.1347e+006
17. set omega [expr pow($K/$M,0.5)]
18. element truss   1       1       2  1.0    1
19. fix 1 1 1
20. fix 2 0 1
21. set accelSeries "Series -dt 0.3 -filePath RV2.txt -factor 1";
22. pattern UniformExcitation 1 1 -accel$accelSeries
23. recorder Node -file nodeDisp_1.out  -time -node 2 -dof 1  disp
24. recorder Element -file axial.out -time -element 1 axialForce
25. recorder  Element  -file  axial.out   -time   -element  1
    axialDeformation
26. # STRUCTURAL ANALYSIS MODEL
27. system BandSPD
28. constraints Plain
29. test NormDispIncr 1.0e-16  50
30. algorithm Newton
31. numberer RCM
32. integrator Newmark 0.5 0.25 [expr 2*0.02*$omega] 0.0  0.0  0.0
33. analysis Transient
34. set totalTime    3.0
35. set numPulses     10
36. set numTimeSteps [expr $numPulses*10]
```

```
37. set phi0 0.25
38. # UNCERTAINTY CHARACTERIZATION
39. set deltat [expr $totalTime/$numPulses]
40. analyze $numTimeSteps
    [expr $totalTime/$numTimeSteps]
41. set u21 [nodeDisp 2 1]
42. set fo [open PerformanceU.txt w] ;
43. foreachdifferent_content {u21} {
44. puts $fo [set $different_content]
45. }
46. close $fo
```

Next, another method of mean upcrossing rates analysis is performed. The same inputs as the previous case are used for performing the reliability analysis, except the "***runOutCrossingAnalysis***" in line 18 is modified as follows:

```
18 runOutCrossingAnalysis OutCross.out -results
   $numTimeSteps $numTimeSteps $numTimeSteps [expr
   $numTimeSteps/$numPulses] -littleDt 1.0e-6 -twoSearches;
```

Tcl commands of the structural response analysis in OpenSees are the same as in the previous section (i.e., named reliability_3_aux.tcl) except the follow six lines are added after line 46 to record velocity responses of the node.

```
1. set V21 [nodeVel 2 1]
2. set foV [open PerformanceV.txt w] ;
3. foreachdifferent_content {V21} {
4. puts $foV [set $different_content]
5. }
6. close $foV
```

After each structural analysis step is finished, displacement and velocity response values used to compute the LSF are recorded in PerformanceU.txt and PerformanceV.txt files, respectively.

References

[1] Haukaas, T. and Der Kiureghian, A. *User's Guide for Reliability and Sensitivity Analysis in OpenSees*. Pacific Earthquake Engineering Research Center, University of California, Berkeley, CA, 2003.

[2] Gu, Q. *Finite Element Response Sensitivity and Reliability Analysis of Soil-Foundation-Structure-Interaction (SFSI) Systems*. University Of California, San Diego, CA, 2008.

[3] McKenna, F., Fenves, G. L., *et al.* OpenSees (Open System for Earthquake Engineering Simulation) Users Manual, Version 1.6.0, 2004.

[4] Wang Z., Macro B, *et al.* Hamiltonian Monte Carlo methods for subset simulation in reliability analysis. *Struct. Saf.* 2019, *72*: 65–73.

Part II

Introduction to OpenSees Programming

Chapter 15

Downloading the Package and Building OpenSees Source Code

15.1 Download the Necessary Installation Packages

15.1.1 *OpenSees source code*

You can download OpenSees source code at https://github.com/OpenSees/OpenSees directly as shown in Fig. 15.1.

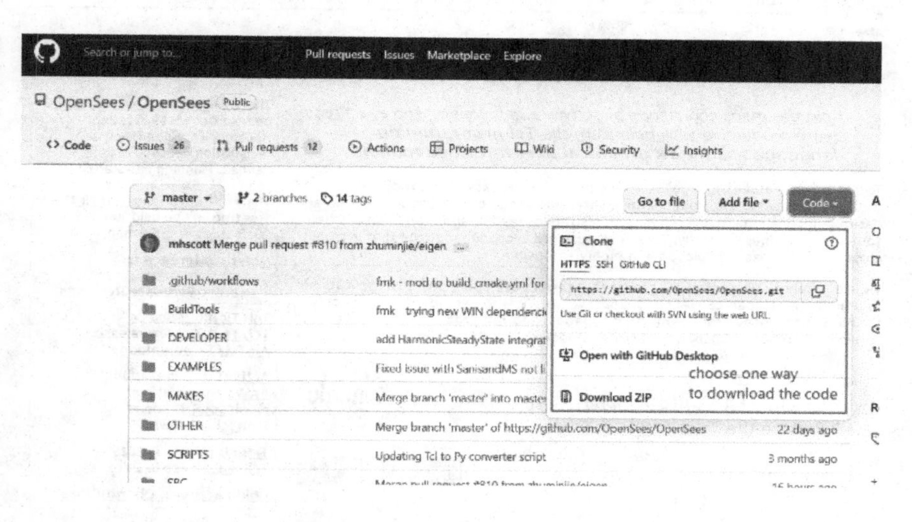

Fig. 15.1 Downloading OpenSees source code on github.

15.1.2 *TCL installation package*

Go to the Tcl official website at https://www.tcl.tk/ to download Tcl (version 8.6). The steps for downloading the Tcl installation package are shown in Figs. 15.2–15.4. After downloading the package, install it into a directory where you can easily find it.

15.1.3 *Python installation package*

If you use python to build models in OpenSees, you need to download and install python (if not, go to Section 15.1.4). Go to the official python website at https://www.python.org/downloads/ and download python (version 3.6 or 3.8). Install it into a directory where you will be able to find it easily. This book focuses only on TCL used to build OpenSees model.

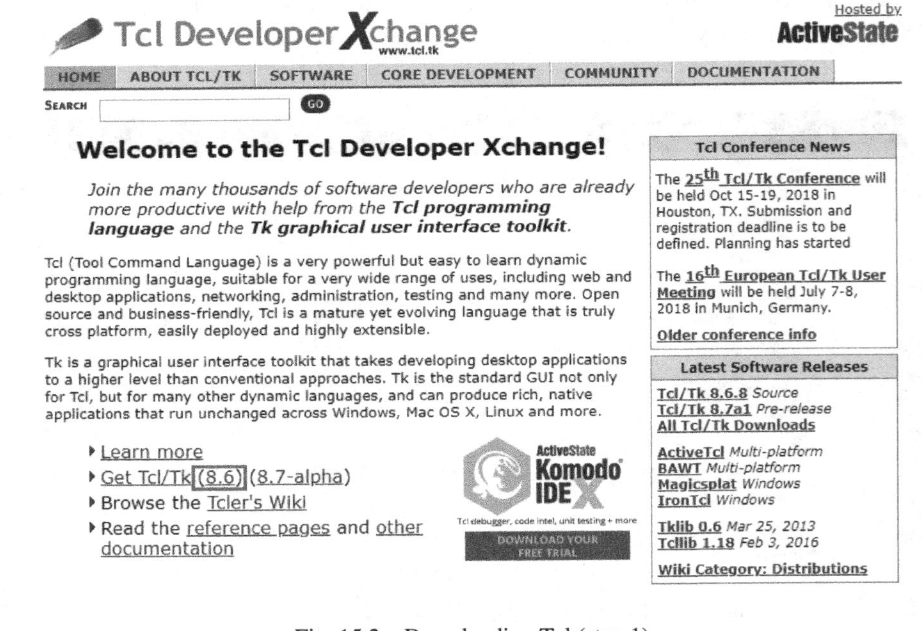

Fig. 15.2 Downloading Tcl (step 1).

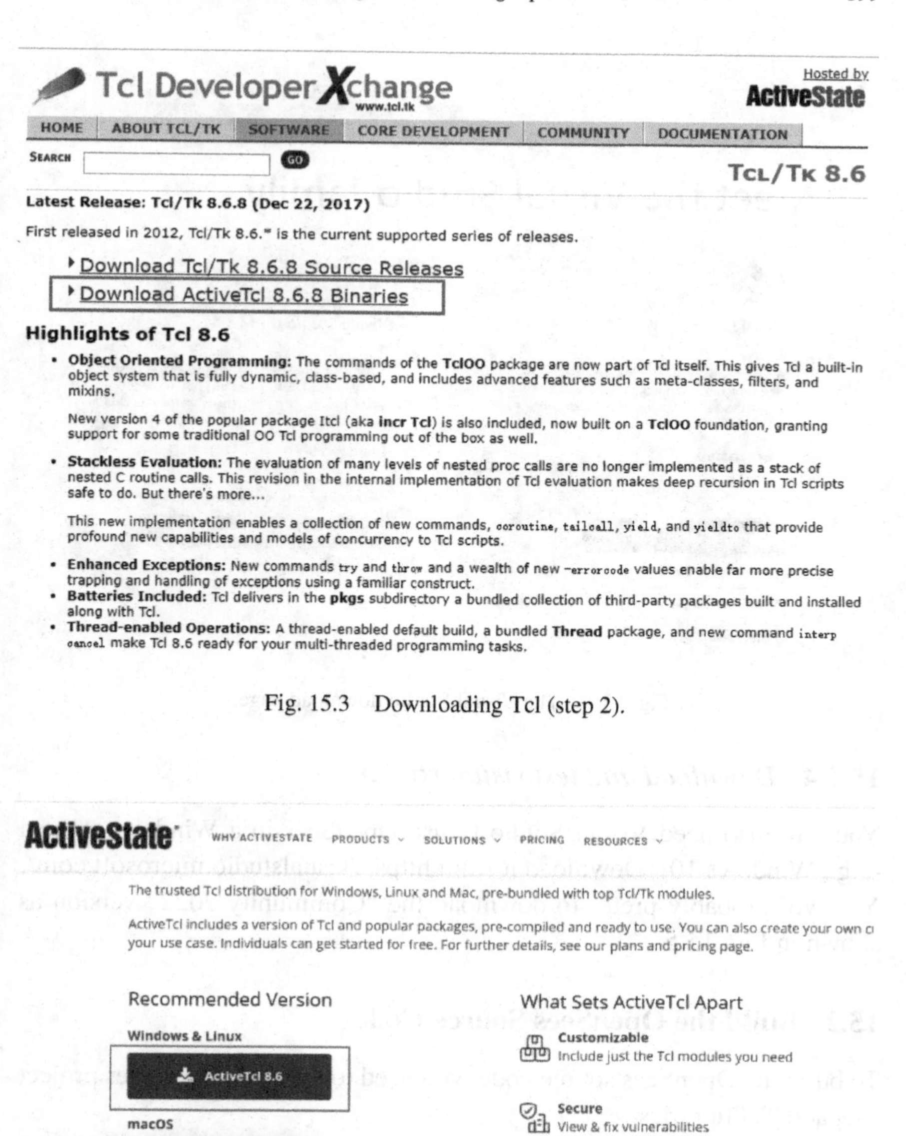

Fig. 15.3 Downloading Tcl (step 2).

Fig. 15.4 Downloading Tcl (step 3).

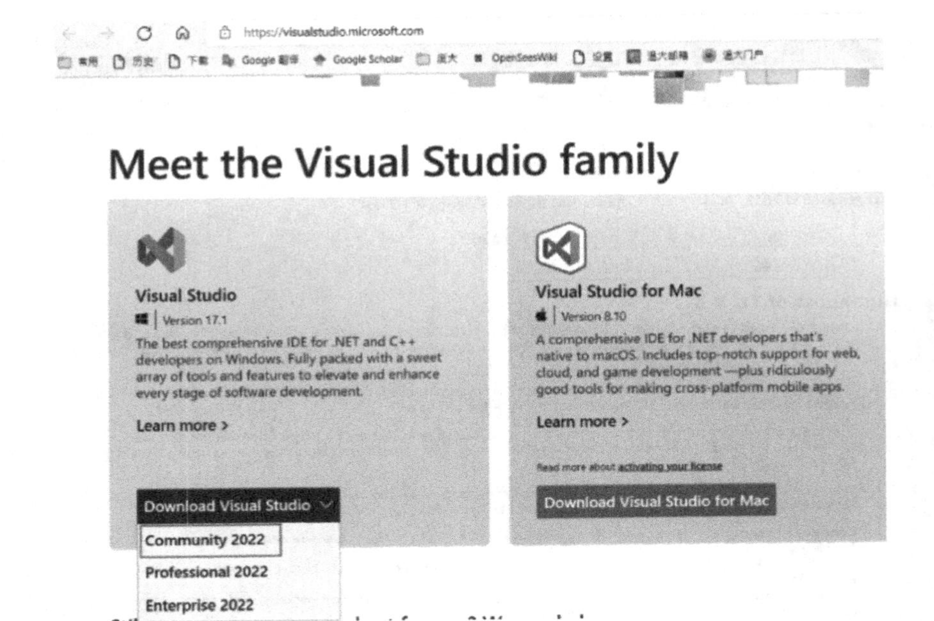

Fig. 15.5 The Visual Studio download page.

15.1.4 *Download and test visual studio*

You will also need Visual Studio to use OpenSees in a Windows system (e.g., Windows 10). Download it from https://visualstudio.microsoft.com/. You will probably prefer to download the "Community 2022" version as shown in Fig. 15.5.

15.2 Build the OpenSees Source Code

To build the OpenSees source code, you need to open the OpenSees project file, add Tcl libraries.

(1) **Open the OpenSees project**
Open Visual Studio 2022, click [Open a project of solution] as shown in Fig. 15.6. Enter the directory of the OpenSees source code you previously download, choose folder "Win32" or "Win64" according to your computer's system operation, select "OpenSees.sln", click [Open] to open the OpenSees project as shown in Fig. 15.7.

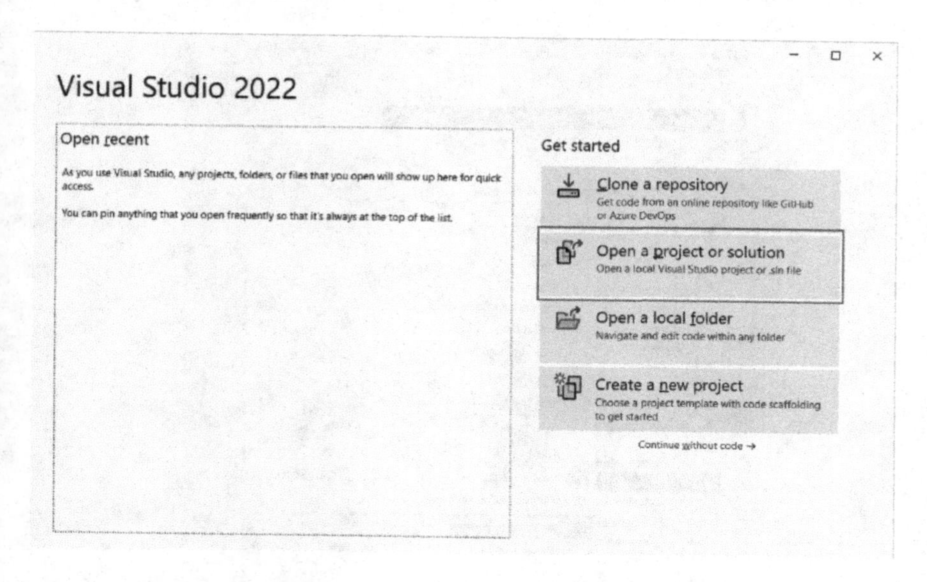

Fig. 15.6 The Visual Studio download page.

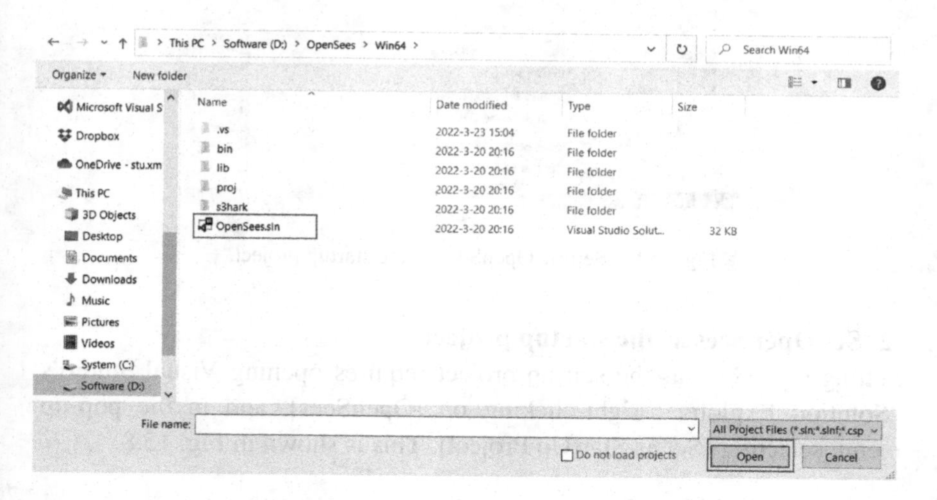

Fig. 15.7 Opening the OpenSees project (2).

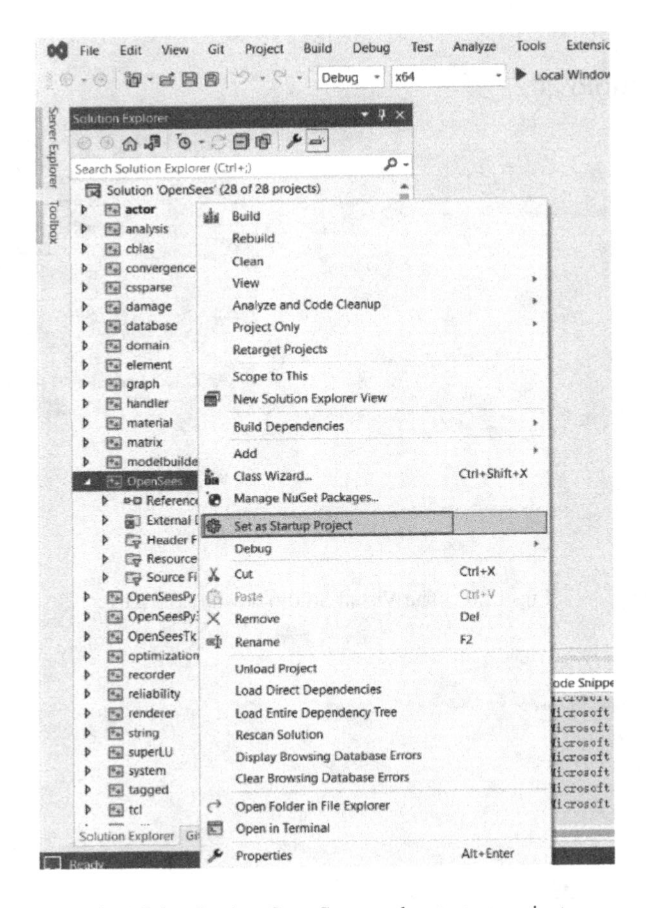

Fig. 15.8 Setting OpenSees as the startup project.

(2) **Set OpenSees as the startup project**

Setting OpenSees as the startup project requires opening Visual Studio's "Solution Explorer", right-clicking on [OpenSees], and in the pop-up menu, selecting [Set as StartUp Project]. This is shown in Fig. 15.8.

(3) **Add the Tcl libraries**

Click [View] → [Property Manager] in the menu bar (see Fig. 15.9) and select [OpenSees] → [Debug | x64], then double-click [Microsoft.Cpp.x64.user] (see Fig. 15.10). In the pop-up window "Microsoft.Cpp.x64.user Property Pages", select [Common Properties] → [VC++ Directories], and

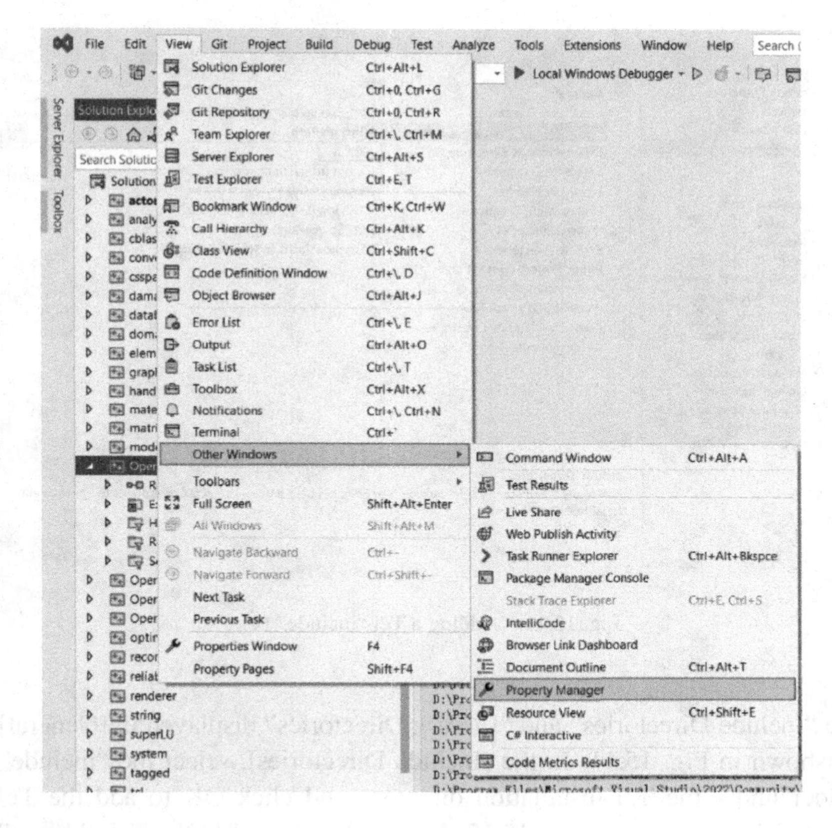

Fig. 15.9 Opening the menu “Property Manager”.

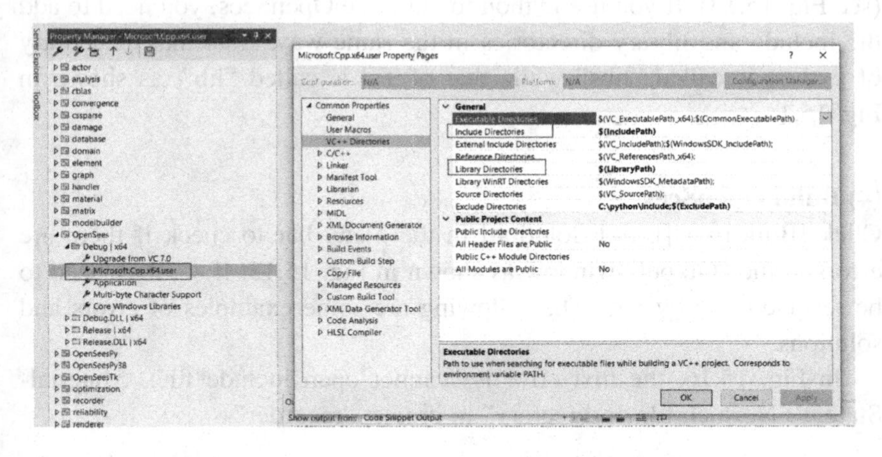

Fig. 15.10 Opening the popup window “Microsoft.Cpp.x64.user Property Pages”.

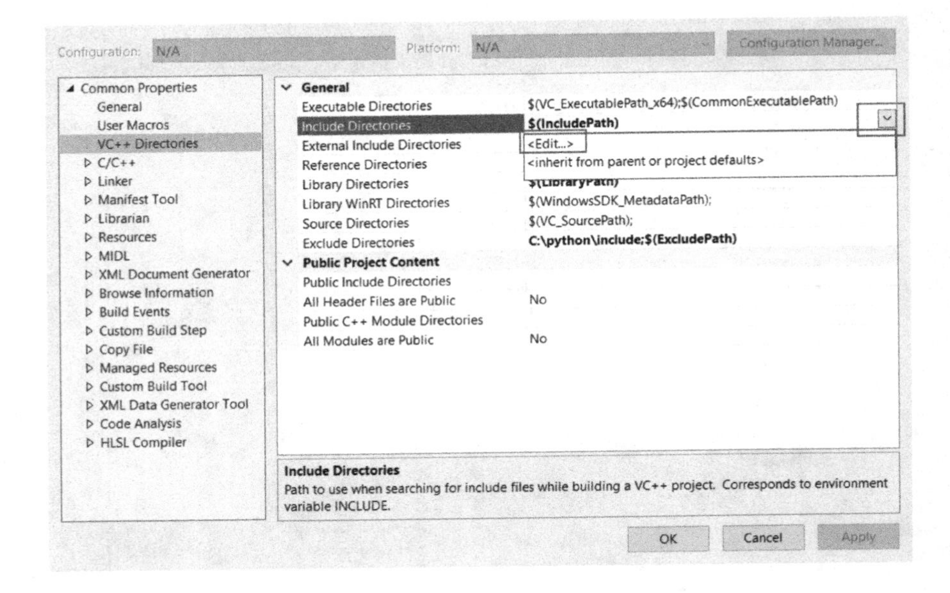

Fig. 15.11 Adding a Tcl "include" (1).

the "Include Directories" and "Library Directories" displayed in [General] as shown in Fig. 15.11. In the [Include Directories], select the "include" folder under the Tcl installation directory and click OK to add the Tcl "include", as shown in Fig. 15.12. In the same way, add the Tcl "library" (see Fig. 15.13). If you use python to build the OpenSees, you need to add the include and library directories in the same way. Note that the library of python is called "libs" while that of Tcl is called "lib", as shown in Fig. 15.14.

(4) **Build OpenSees**

Click [Build] → [Build Solution] in the menu bar to check if there are errors in the "Output" window as shown in Fig. 15.15. If so, they need to be resolved one by one. The following are some examples of errors and solutions.

In Fig. 15.16, the first error is "cannot open include file: 'Material-State.h': No such file or directory" in project "recorder".

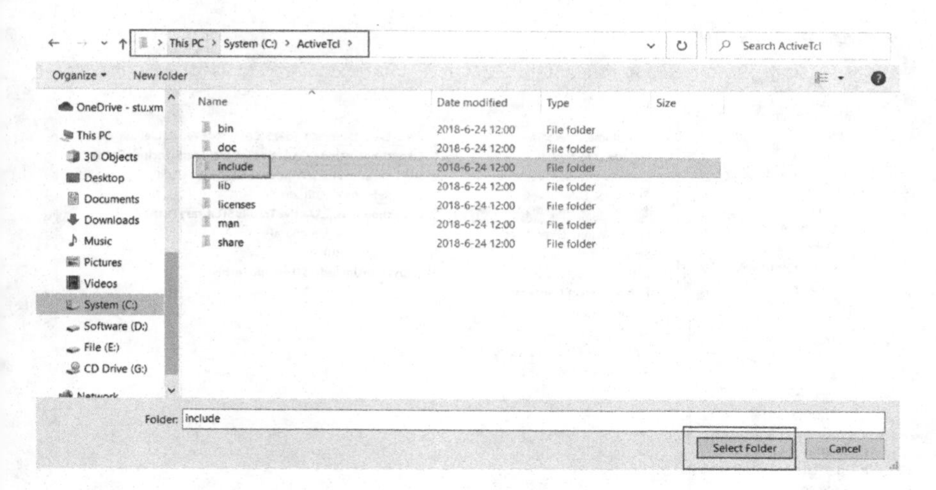

Fig. 15.12 Adding a Tcl "include" (2).

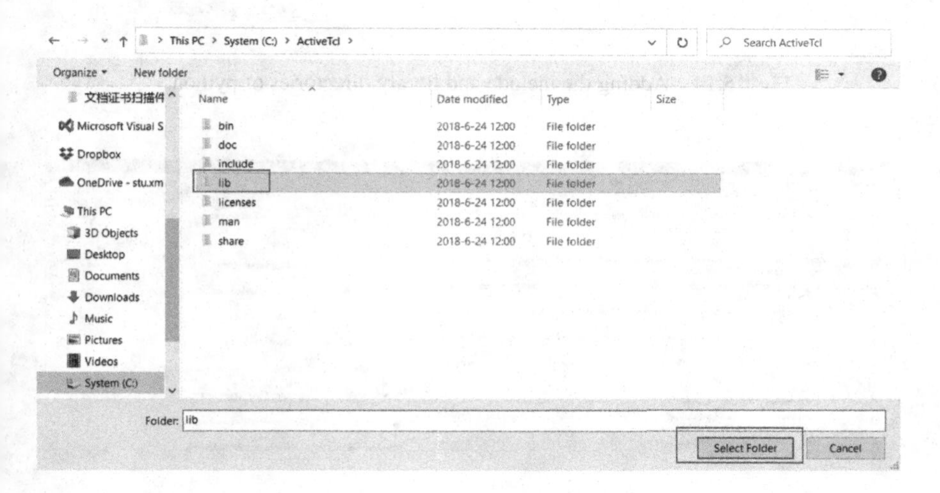

Fig. 15.13 Adding a Tcl "library".

One way to solve this error is to find the location of MaterialState.h: search for the file in [Solution Explorer], and it is found in the folder [material\uniaxial]. Second, double-click [recorder], in the pop-up menu, click [Properties] (see Fig. 15.17), select [C/C++] → [General] → [Additional Include Directories], and add the path

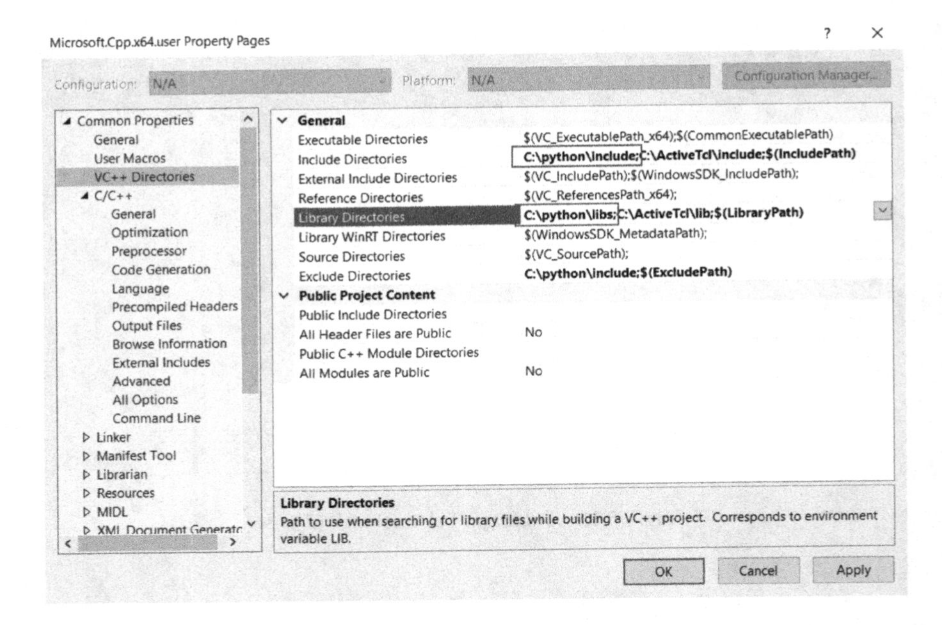

Fig. 15.14 Adding the include and library directories of python.

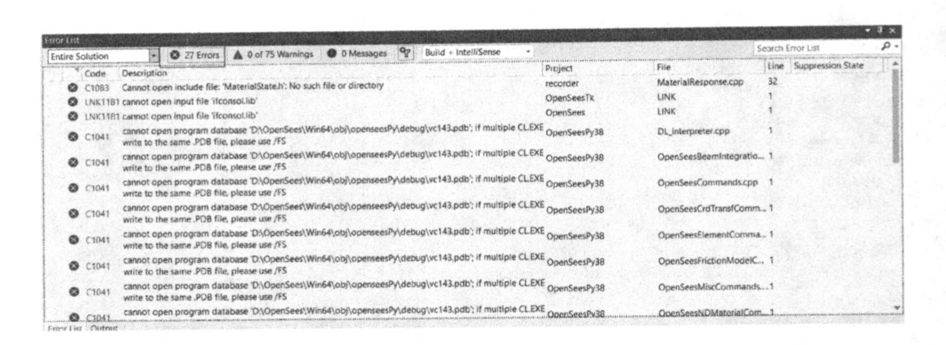

Fig. 15.15 Errors in building OpenSees.

"..\..\..\src\materal\uniaxial" (see Fig. 15.18), click [OK,] and build OpenSees again, you will find that the error in project "recorder" is solved, as shown in Fig. 15.19.

Figure 15.19 shows some new errors. You may find the errors are all in projects "OpenSeesPy" and "OpenSeesPy38". The easiest way

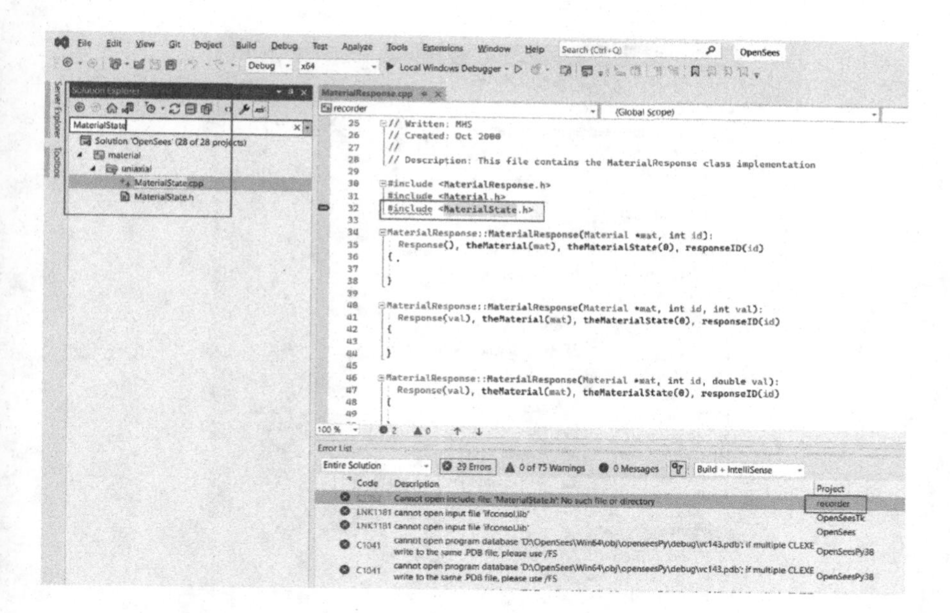

Fig. 15.16 An error in project "recorder".

to solve the problems is to click [Build] → [Configuration Manager] (see Fig. 15.20), cancel the "OpenSeesPy" and "OpenSeesPy38" projects and build OpenSees again as shown in Fig. 15.21. However, if you want to use the python project, you need to solve the problems one by one.

Build OpenSees again, and you will find some new errors as shown in Fig. 15.22. To solve these errors, you need to copy the .lib files to the folder in which you have installed OpenSees, i.e., .\OpenSees\Win64\lib see Fig. 15.23. If you do not have the files, you may download them at https://github.com/OpenSeesXMU/OpenSees/tree/master/Win64/lib.

Build OpenSees again, an error appears "cannot open input file 'PML.lib'", as shown in Fig. 15.24. A simple way to solve the problem is proposed here if you do not need to use the element "PML2D" or "PML3D" (see Figs. 15.25–15.27): first, double-click [OpenSees], select [Properties], in the pop-up menu, click [Linker] -> [Input] -> [Additional Dependencies], find PML.lib, and delete it. Do the same for the project "OpenSeesTk". Second, find "PML2D.cpp" and "PML3D.cpp" in element/PML, and comment on the codes as shown in Fig. 15.27. If you need to use these elements, you may find other ways to solve the errors.

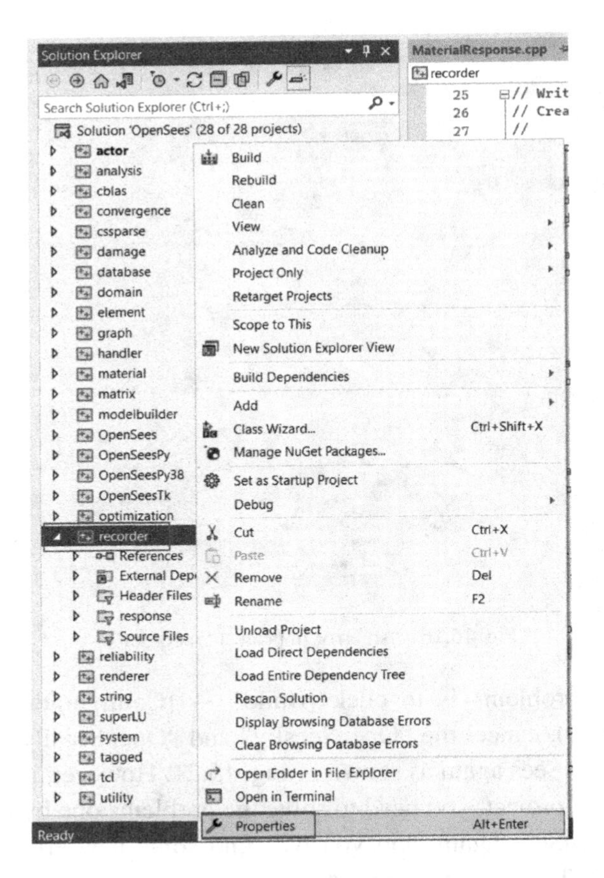

Fig. 15.17 Solve the error in project “recorder” (1).

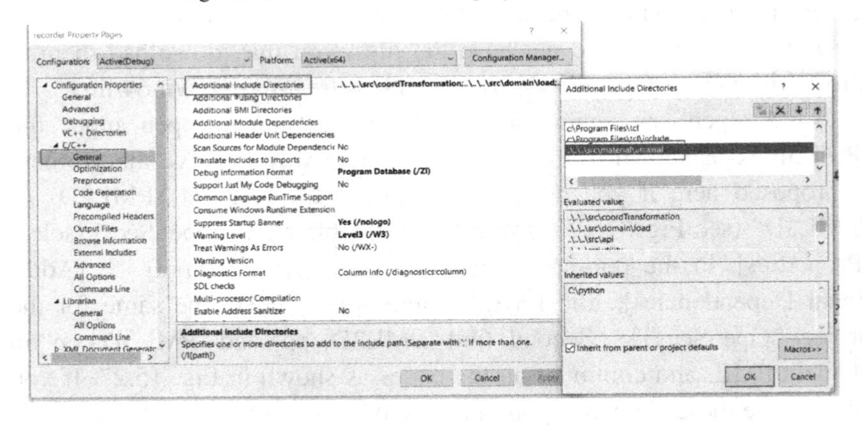

Fig. 15.18 Solve the error in project “recorder” (2).

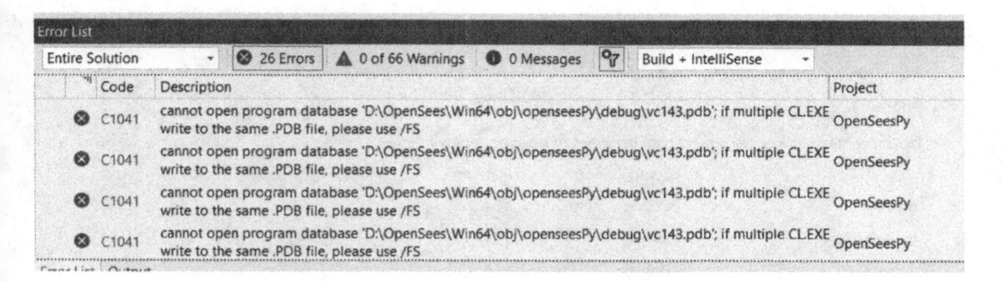

Fig. 15.19 Solve the error in project “recorder” (3).

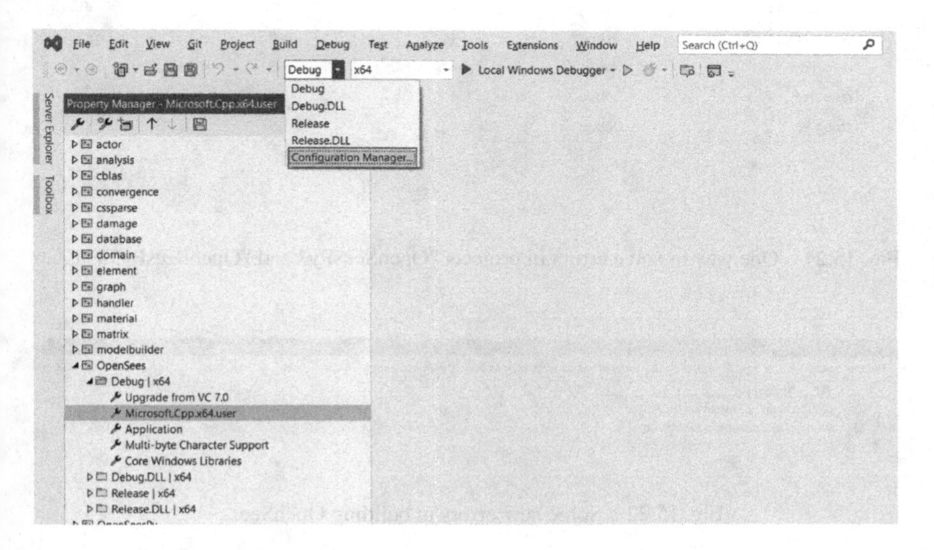

Fig. 15.20 One way to solve errors in projects “OpenSeesPy” and “OpenSeesPy38” (1).

Build OpenSees again, all the errors are solved, and OpenSees.exe and OpenSeesTk.exe are successfully built where you can find them in the folder \OpenSees\Win64\bin (see Fig. 15.28). Click [Debug] -> [Start Debugging], or press F5, the debug mode runs as shown in Fig. 15.29.

If you still have troubles building OpenSees, you can download the repository at https://github.com/OpenSeesXMU/OpenSees, which can be built directly using Visual Studio 2017 (or later version), with Tcl 8.6 in the Windows system.

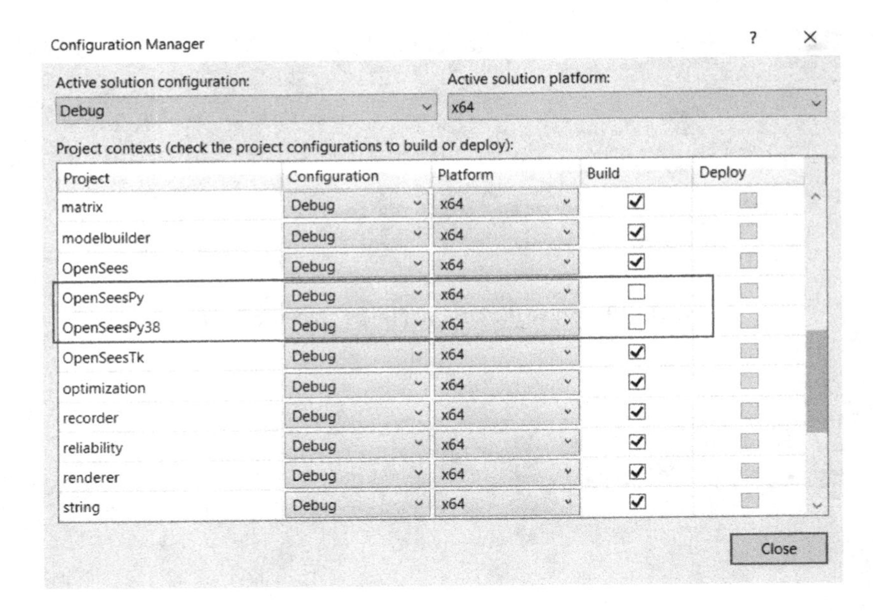

Fig. 15.21 One way to solve errors in projects “OpenSeesPy” and “OpenSeesPy38” (2).

Fig. 15.22 Some new errors in building OpenSees.

Fig. 15.23 Copy the files.

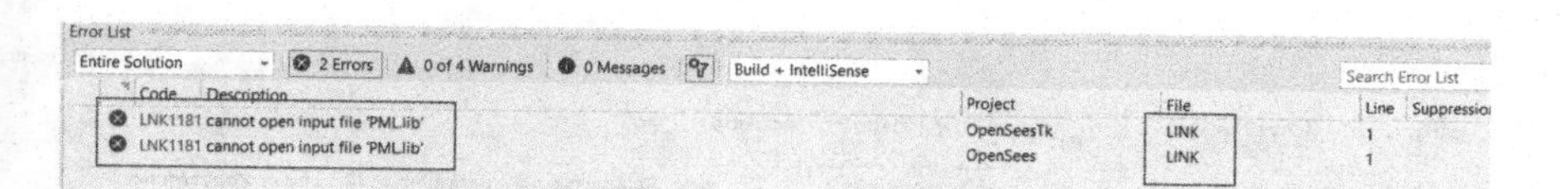

Fig. 15.24 New errors in building OpenSees.

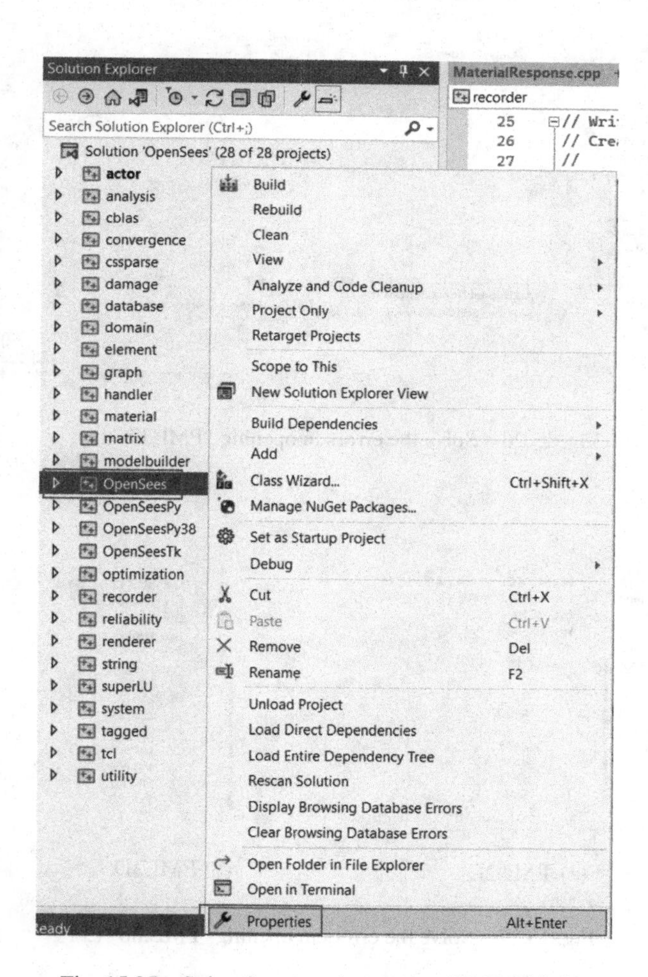

Fig. 15.25 Solve the errors in opening "PML.lib" (1).

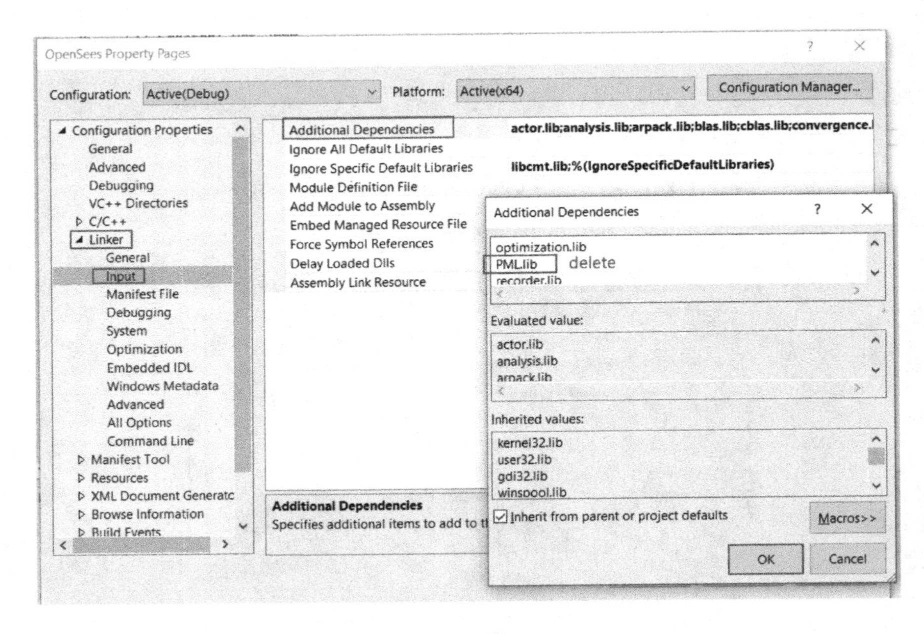

Fig. 15.26 Solve the errors in opening "PML.lib" (2).

```
142
143     int NDOFEL = PML2D_NUM_DOF;
144     int NPROPS = PML2D_NUM_PROPS;
145     int MCRD = 2;
146     int NNODE = PML2D_NUM_NODES;
147
148  //pml2d_(K,
149  //      C,
150  //      M,
151  //      &NDOFEL,
152  //      props,
153  //      &NPROPS,
154  //      coords,
155  //      &MCRD,
156  //      &NNODE);
157  }
158
```

(a) PML2D

```
388
389     int NDOFEL = PML3D_NUM_D
390     int NPROPS = 12;
391     int MCRD = 3;
392     int NNODE = 8;
393
394     //pml3d_(M,
395     // C,
396     // K,
397     // &NDOFEL,
398     // props,
399     // &NPROPS,
400     // coords,
401     // &MCRD,
402     // &NNODE);
403
```

(b) PML3D

Fig. 15.27 Solve the errors in opening "PML.lib" (3).

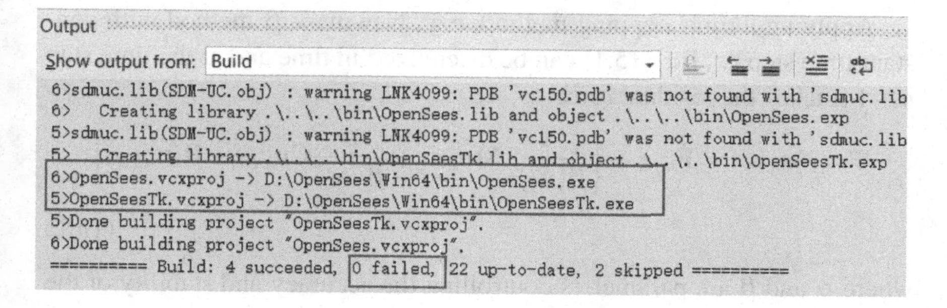

Fig. 15.28 Solve the errors in opening "PML.lib" (1).

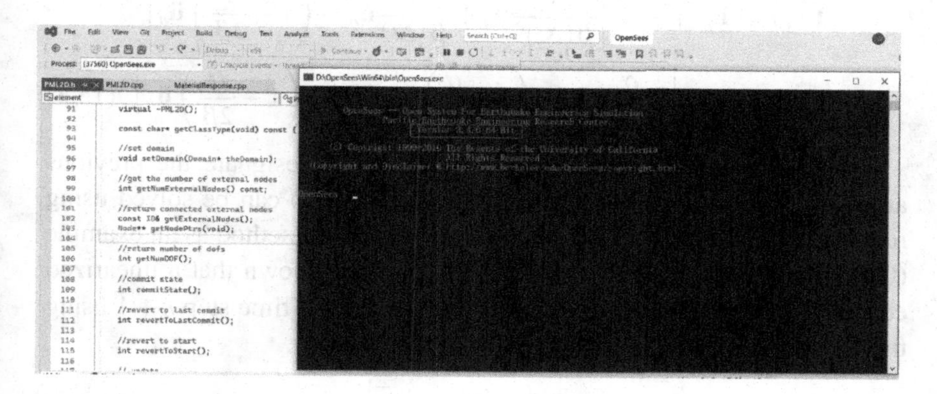

Fig. 15.29 The debug mode of OpenSees.

15.3 A Typical Flowchart in OpenSees

The flowchart in OpenSees is complex. For illustration purposes, a typical flowchart in OpenSees is introduced when implicit time stepping method and Newton Raphson solution algorithm are applied. The governing equilibrium equation of motion can be expressed as

$$\mathbf{M}\ddot{\mathbf{u}} + \mathbf{C}\dot{\mathbf{u}} + \mathbf{R}(\mathbf{u}) = \mathbf{F} \tag{15.1}$$

where **M** and **C** are the mass and damping matrices. **u**, **R**, and **F** are the vectors of nodal displacement, the internal resisting force, and applied dynamic force of the system, respectively. The superposed dot indicates differentiation with respect to time. $\dot{\mathbf{u}}$ and $\ddot{\mathbf{u}}$ are the nodal velocity vector and the nodal acceleration vector, respectively.

Applying a time stepping method, e.g., Newmark-β method with constant time step Δt, Eq. (15.1) can be discretized in time and at the time step $n+1$ it takes the residual form

$$\Psi_{n+1}(\mathbf{u}_{n+1}) = \tilde{\mathbf{F}}_{n+1} - \left[\frac{1}{\beta(\Delta t)^2}\mathbf{M}\mathbf{u}_{n+1} + \frac{\alpha}{\beta(\Delta t)}\mathbf{C}\mathbf{u}_{n+1} + \mathbf{R}(\mathbf{u}_{n+1})\right] \tag{15.2}$$

where α and β are parameters controlling the accuracy and stability of the numerical integration algorithm and

$$\begin{aligned}\tilde{\mathbf{F}}_{n+1} = \mathbf{F}_{n+1} &+ \mathbf{M}\left[\frac{1}{\beta(\Delta t)^2}\mathbf{u}_n + \frac{1}{\beta(\Delta t)}\dot{\mathbf{u}}_n - \left(1 - \frac{1}{2\beta}\right)\ddot{\mathbf{u}}_n\right] \\ &+ \mathbf{C}\left[\frac{\alpha}{\beta(\Delta t)}\mathbf{u}_n - \left(1 - \frac{\alpha}{\beta}\right)\dot{\mathbf{u}}_n - (\Delta t)\left(1 - \frac{\alpha}{2\beta}\right)\ddot{\mathbf{u}}_n\right]\end{aligned} \tag{15.3}$$

In general, the subscripts $(\cdots)_n$ and $(\cdots)_{n+1}$ designate the previous and current time steps, respectively. Equation (15.3) can be solved using root-finding algorithms. Take the Newton–Raphson method as an example (the principle can be seen in Fig. 15.30) [1], it is shown that a linearized equation needs to be solved at each iteration i of the time step $n+1$ using the Newton–Raphson method, i.e.,

$$\Delta\mathbf{u}^i_{n+1} = \left(-\frac{\partial\Psi^i_{n+1}(\mathbf{u}^i_{n+1})}{\partial\mathbf{u}^i_{n+1}}\right)^{-1}\Psi^i_{n+1}(\mathbf{u}^i_{n+1}) \tag{15.4}$$

The nodal displacement after the ith iteration is updated as

$$\mathbf{u}^{i+1}_{n+1} = \mathbf{u}^i_{n+1} + \Delta\mathbf{u}^i_{n+1} \tag{15.5}$$

In each iteration, the displacement is called "trial displacement", i.e., $\mathbf{u}^0_{n+1}, \mathbf{u}^1_{n+1}, \mathbf{u}^2_{n+1}$ and $\mathbf{u}^i_{n+1}$ as shown in Fig. 15.30. At the end of each iteration, it should be checked whether or not a convergence criterion has been satisfied, e.g., norm of the displacement increments, norm of the unbalanced forces or energy imbalance. If norm of unbalance forces is used, then

$$||\Psi_{n+1}(\mathbf{u}_{n+1})|| \leq \text{tolerance} \tag{15.6}$$

If the convergence criterion is satisfied, then the current step is committed (i.e., set trial variables to committed variables), as shown in Fig. 15.30.

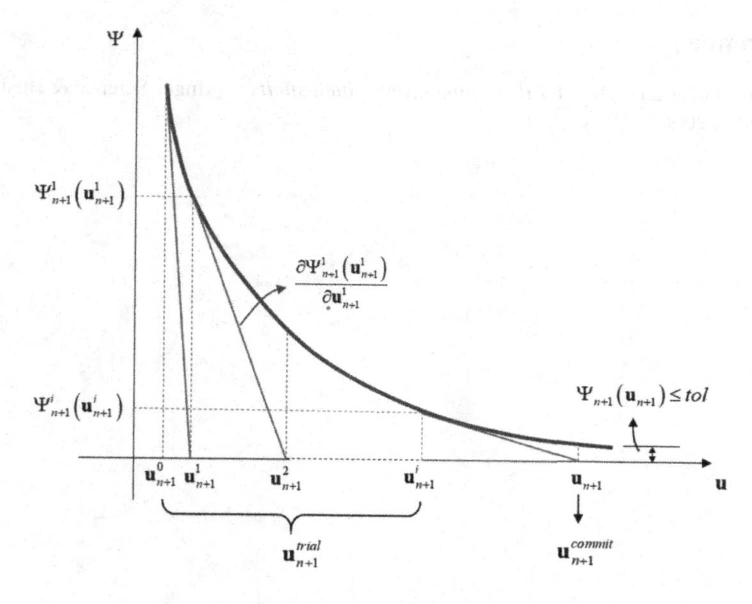

Fig. 15.30 The Newton–Raphson method.

The above procedure in OpenSees is summarized as follows:

Loop each time step n

Step 1. Apply the external force at current time step $n+1$

Step 2. Solve the equation of motion, i.e., Eq. (15.3), using the Newton method

2.1 Form unbalance force $\Psi_{n+1}(\mathbf{u}_{n+1})$ in Eq. (15.3)

2.2 Check whether the convergence criterion of Eq. (15.6) is satisfied.
IF YES, go to Step 3; ELSE

2.3 Form the tangent stiffness matrix, $\frac{\partial \Psi^i_{n+1}(\mathbf{u}^i_{n+1})}{\partial \mathbf{u}^i_{n+1}}$ using Eq. (15.4)

2.4 Calculate $\Delta \mathbf{u}^i_{n+1}$ using Eq. (15.4)

2.5 Update $\mathbf{u}^i_{n+1}$ using Eq. (15.5)

2.6 go to Step 2.1

Step 3. Commit the current step

End loop

Reference

[1] Simo J C and Hughes T J R. *Computational Inelasticity*. Springer Science & Business Media, 2006.

Chapter 16

Adding an Elasto-Perfect-Plasticity Material in OpenSees

16.1 Brief Introduction

This chapter introduces how to add an elasto-perfect-plasticity material model in OpenSees. The computer's operation system is Windows 10 and the IDE is Visual Studio 2017 (while the codes in the later versions are similar). The new material is named "PerfectPlasticMaterial".

The new material "PerfectPlasticMaterial" class inherits the one-dimensional material class "UniaxialMaterial" as its base class. The constitutive behavior of this model is shown in Fig. 16.1. To implement it, a few crucial member functions will be required, including a constructor function, setTrialStrain(), commitState (), etc.

16.2 Defining the Material

The following steps are needed to add a new material:

Step 1: Find the uniaxial material "NewUniaxialMaterial" (including ".cpp" and ".h" files) in a folder named "\src\material\uniaxial" in OpenSees folder. Copy the two files and paste them into the same folder, then rename them as "PerfectPlasticMaterial.h" and "PerfectPlasticMaterial.cpp", respectively, as shown in Fig. 16.2.

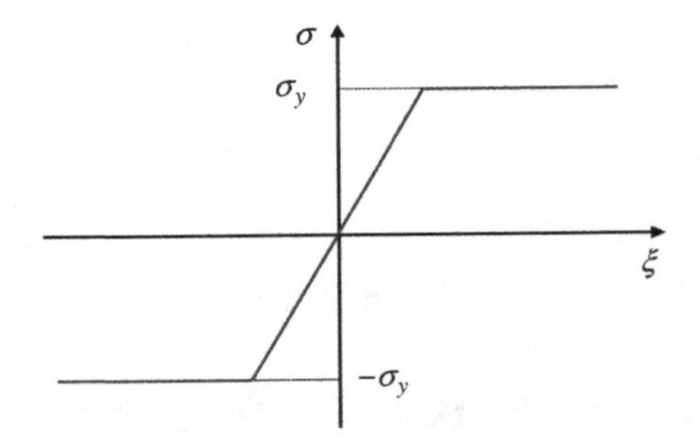

Fig. 16.1 The constitutive behavior of the perfectly elastic plastic material.

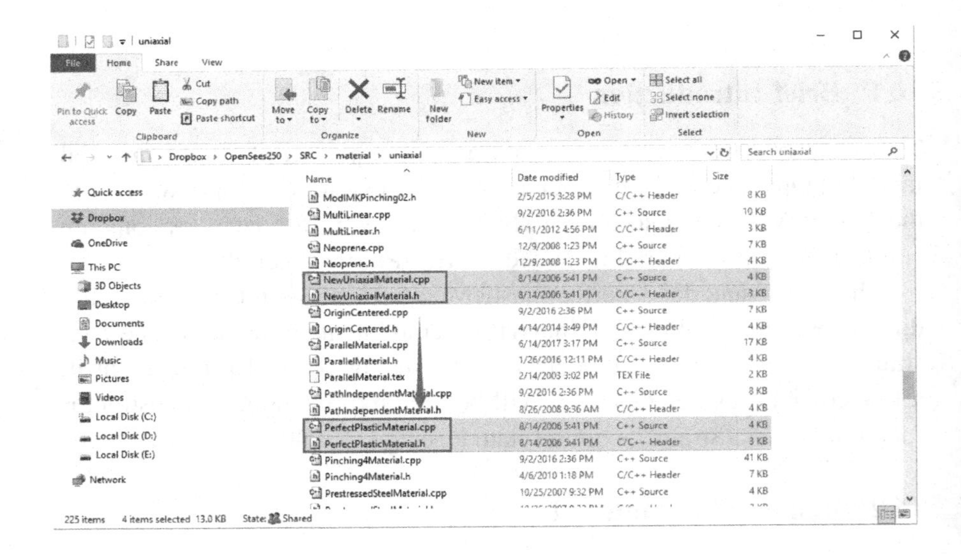

Fig. 16.2 Copying and renaming the files.

Step 2: Add “PerfectPlasticMaterial.h” and “PerfectPlastic-Material.cpp” to the “Solution Explorer” as shown in Figs. 16.3 and 16.4. Rebuild the source code and confirm that it is successful at this step (it is suggested to compile the code from time to time once a relatively “complete” piece of code is done, which is useful for debugging a very long code).

Step 3: Open the “PerfectPlasticMaterial.h” and “PerfectPlastic-Material.cpp” files. Replace all the words “NewUniaxialMaterial” with

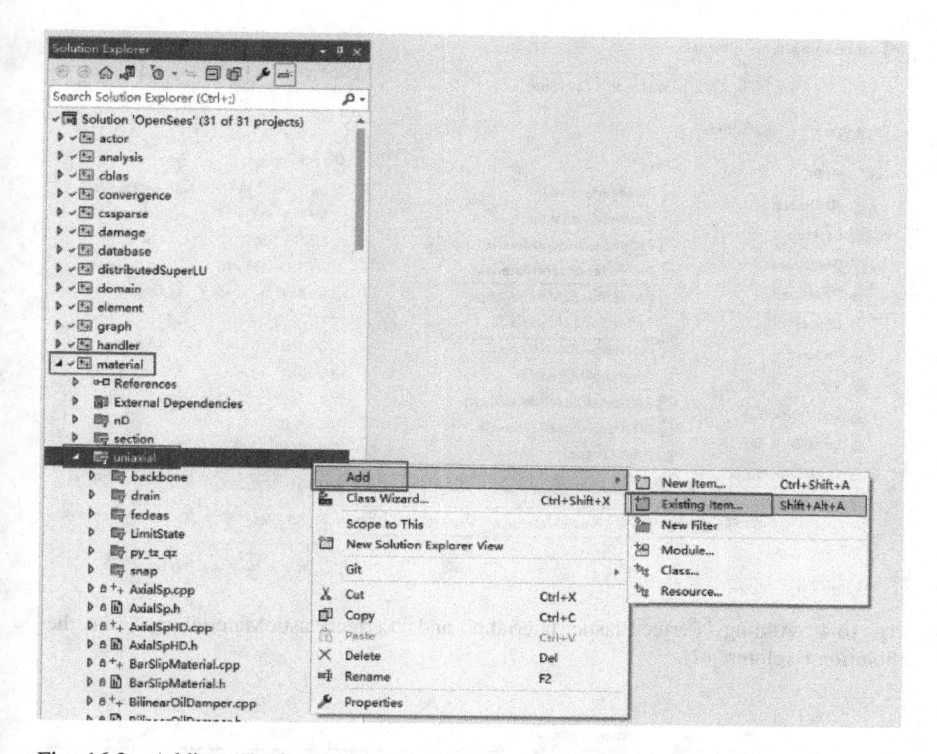

Fig. 16.3 Adding "PerfectPlasticMaterial.h" and "PerfectPlasticMaterial.cpp" into the "Solution Explorer" (1).

"PerfectPlasticMaterial" in the two files, as shown in Fig. 16.5–16.8. Assign a new and unique integer material identifier (e.g., 20180610 in this example) to "MAT_TAG_NewUniaxialMaterial" as shown in Figs. 16.9 and 16.10. Finally, re-build the source code and make sure that there were no errors in this step.

Step 5: Add and modify some private members in the file "PerfectPlastic-Material.h" as shown in Fig. 16.11.

Step 6: Add and modify some definitions of the members' functions in the file "PerfectPlasticMaterial.h" as shown in Fig. 16.12.

Step 7: Implement in the ".cpp" file the member functions declared in the ".h" file. Make sure that there are no errors by clicking [Build] to build

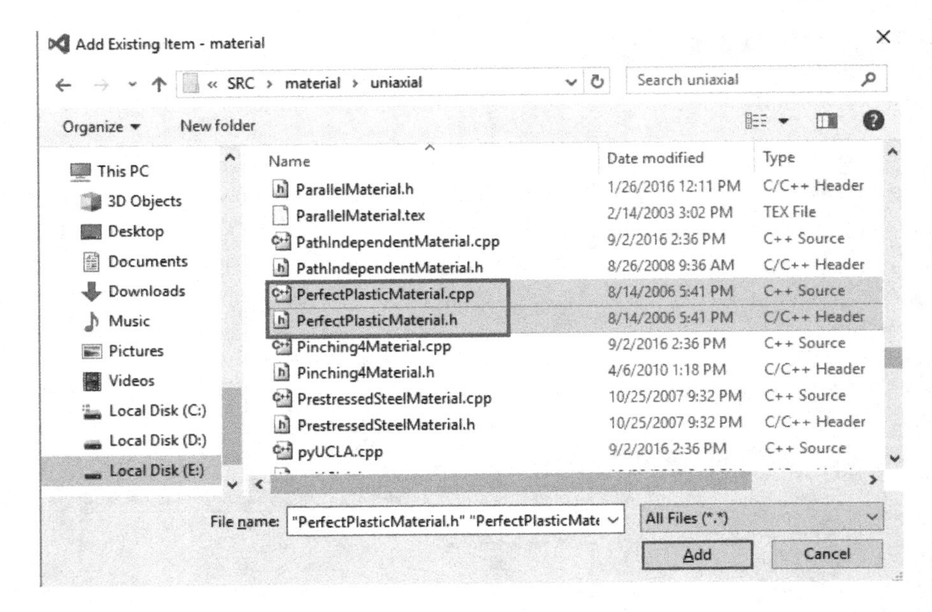

Fig. 16.4 Adding "PerfectPlasticMaterial.h" and "PerfectPlasticMaterial.cpp" into the "Solution Explorer" (2).

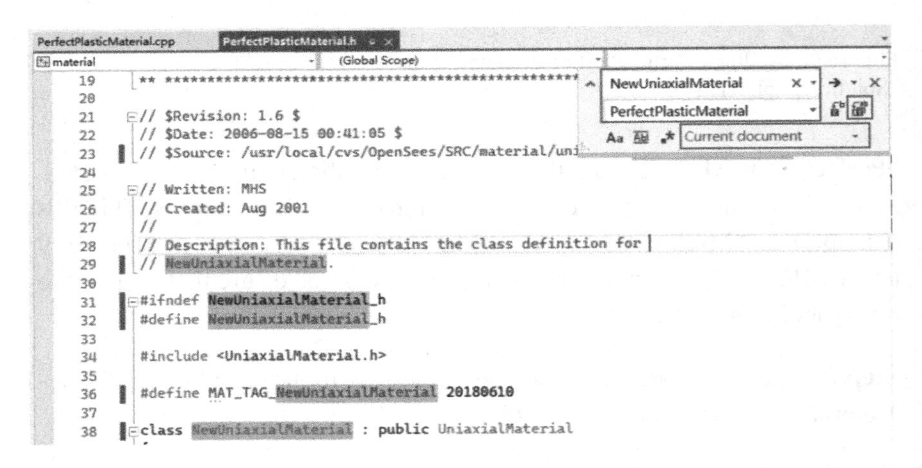

Fig. 16.5 Replacing all the occurrences of "NewUniaxialMaterial" with "PerfectPlastic-Material" in "PerfectPlasticMaterial.h" (1).

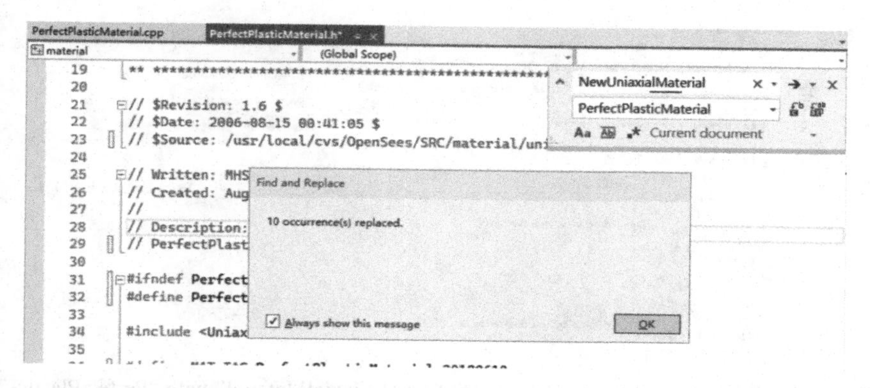

Fig. 16.6 Replacing all the occurrences of "NewUniaxialMaterial" with "PerfectPlastic-Material" in "PerfectPlasticMaterial.h" (2).

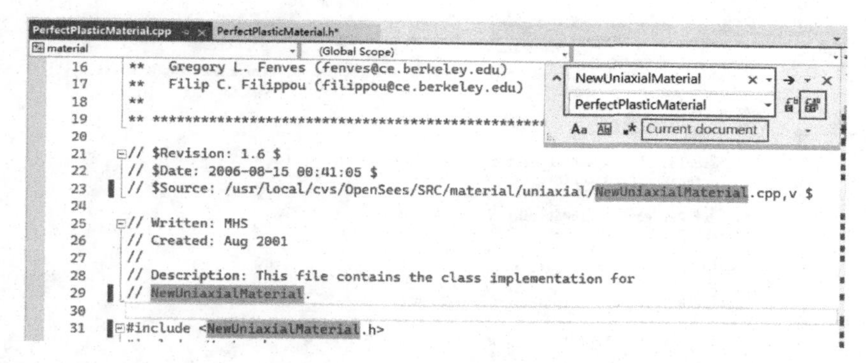

Fig. 16.7 Replacing all the occurrences of "NewUniaxialMaterial" with "PerfectPlastic-Material" in "PerfectPlasticMaterial.cpp" (1).

the newly-added source code. In this example, the member functions are modified as follows:

♦ PerfectPlasticMaterial()

```
1. PerfectPlasticMaterial::PerfectPlasticMaterial(int tag,double pE,
   double pStress0)
2. :UniaxialMaterial(tag,MAT_TAG_PerfectPlasticMaterial),
3. {
4. E = pE;
5. Stress0 = pStress0;
6. this->revertToStart();
7. }
```

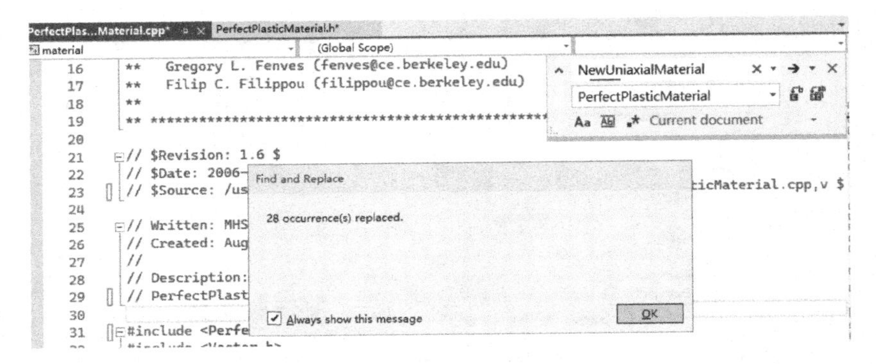

Fig. 16.8 Replacing all the occurrences of “NewUniaxialMaterial” with “PerfectPlastic-Material” in “PerfectPlasticMaterial.cpp” (2).

```
23  // $Source: /usr/local/cvs/OpenSees/SRC/material/uniaxial/PerfectPlasticMaterial.h,v $
24
25  // Written: MHS
26  // Created: Aug 2001
27  //
28  // Description: This file contains the class definition for
29  // PerfectPlasticMaterial.
30
31  #ifndef PerfectPlasticMaterial_h
32  #define PerfectPlasticMaterial_h
33
34  #include <UniaxialMaterial.h>
35
36  #define MAT_TAG_PerfectPlasticMaterial 1976
37
38  class PerfectPlasticMaterial : public UniaxialMaterial
39  {
40   public:
```

Fig. 16.9 Assigning a new integer value to “MAT_TAG_NewUniaxialMaterial” (1).

♦ setTrialStrain()

```
1. int
2. PerfectPlasticMaterial::setTrialStrain(double strain, double
   strainRate)
3. {
4. // elastic predictor
5. trialStrain = strain;
6. double dStrain = trialStrain - CStrain;
7. trialStress = CStress + E*dStrain;
8. trialTangent = E;
```

9. // plastic corrector
10. double eps = 1e-14;
11. if(trialStress > Stress0)
    a) {trialStress = Stress0;trialTangent = eps;}
12. if(trialStress < -Stress0)
    a) {trialStress = -Stress0;trialTangent = eps;}
13. return 0;
14. }

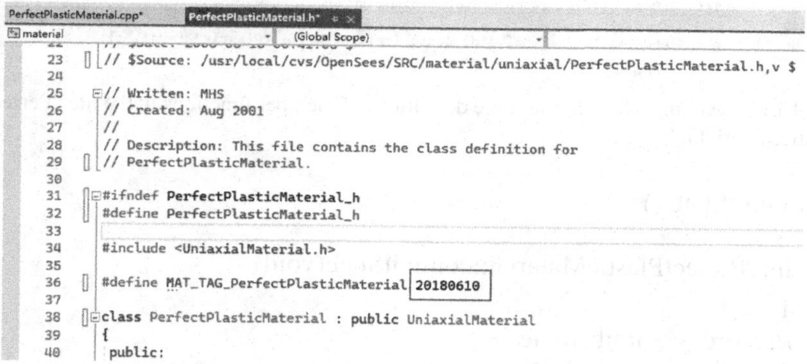

```
PerfectPlasticMaterial.cpp*    PerfectPlasticMaterial.h*
material    (Global Scope)
23  // $Source: /usr/local/cvs/OpenSees/SRC/material/uniaxial/PerfectPlasticMaterial.h,v $
24
25  // Written: MHS
26  // Created: Aug 2001
27  //
28  // Description: This file contains the class definition for
29  // PerfectPlasticMaterial.
30
31  #ifndef PerfectPlasticMaterial_h
32  #define PerfectPlasticMaterial_h
33
34  #include <UniaxialMaterial.h>
35
36  #define MAT_TAG_PerfectPlasticMaterial 20180610
37
38  class PerfectPlasticMaterial : public UniaxialMaterial
39  {
40   public:
```

Fig. 16.10 Assigning a new integer value to "MAT_TAG_NewUniaxialMaterial" (2).

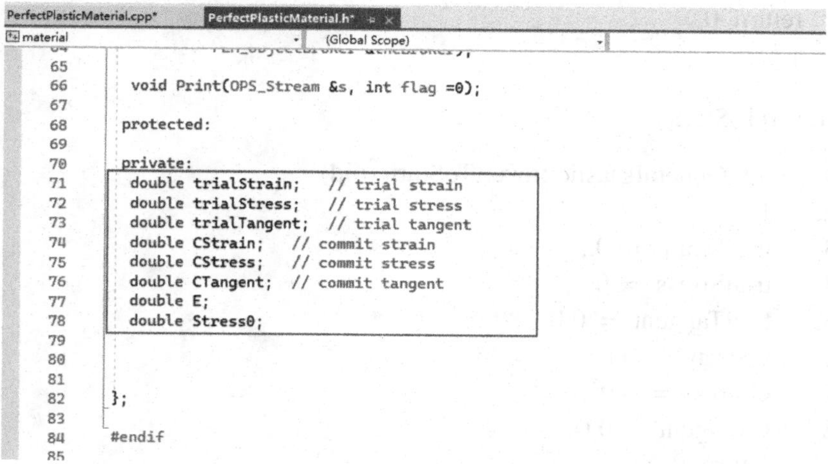

```
PerfectPlasticMaterial.cpp*    PerfectPlasticMaterial.h*
material    (Global Scope)
65
66    void Print(OPS_Stream &s, int flag =0);
67
68   protected:
69
70   private:
71    double trialStrain;   // trial strain
72    double trialStress;   // trial stress
73    double trialTangent;  // trial tangent
74    double CStrain;   // commit strain
75    double CStress;   // commit stress
76    double CTangent;  // commit tangent
77    double E;
78    double Stress0;
79
80
81
82  };
83
84  #endif
85
```

Fig. 16.11 Adding and modifying some private members in the file "PerfectPlasticMaterial.h".

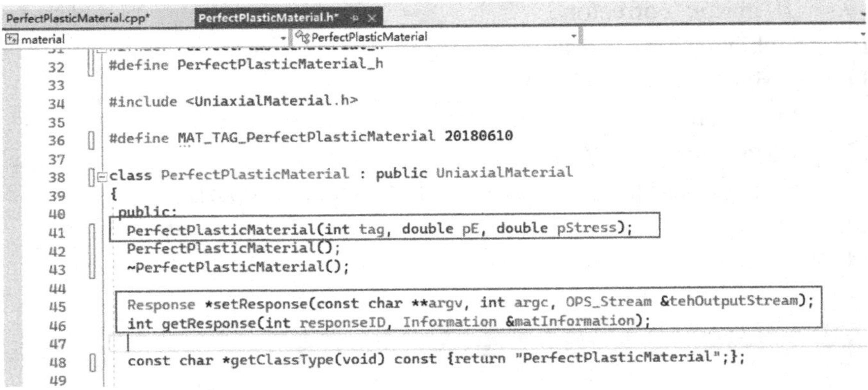

```
PerfectPlasticMaterial.cpp*   PerfectPlasticMaterial.h*
material                      PerfectPlasticMaterial
32  #define PerfectPlasticMaterial_h
33
34  #include <UniaxialMaterial.h>
35
36  #define MAT_TAG_PerfectPlasticMaterial 20180610
37
38  class PerfectPlasticMaterial : public UniaxialMaterial
39  {
40   public:
41    PerfectPlasticMaterial(int tag, double pE, double pStress);
42    PerfectPlasticMaterial();
43    ~PerfectPlasticMaterial();
44
45    Response *setResponse(const char **argv, int argc, OPS_Stream &tehOutputStream);
46    int getResponse(int responseID, Information &matInformation);
47
48    const char *getClassType(void) const {return "PerfectPlasticMaterial";};
49
```

Fig. 16.12 Adding and deleting some definitions of member functions in the file "PerfectPlasticMaterial.h".

♦ commitState ()

```
1. int PerfectPlasticMaterial::commitState(void)
2. {
3. //record commit value
4. CStrain = trialStrain;
5. CStress = trialStress;
6. CTangent = trialTangent;
7. return 0;
8. }
```

♦ revertToStart()

```
1.  int GeneralElastic::revertToStart(void)
2.  {
3.  trialStrain = 0.;
4.  trialStress = 0.0;
5.  trialTangent = 0.0;
6.  CStrain = 0.;
7.  CStress = 0.0;
8.  CTangent = 0.0;
9.  return 0;
10. }
```

♦ getCopy()

```
1.  UniaxialMaterial * PerfectPlasticMaterial::getCopy(void)
2.  {
3.  PerfectPlasticMaterial *theCopy =
4.  new PerfectPlasticMaterial(this->getTag(),E,Stress0);
5.  return theCopy;
6.  }
```

♦ setResponse()

```
1.  Response*   PerfectPlasticMaterial::setResponse(const   char
    **argv,
2.  int argc, OPS_Stream &theOutput)
3.  {
4.  Response *res = UniaxialMaterial::setResponse(argv, argc,
5.  theOutput);
6.  if (res != 0)  return res;
7.  else           return 0;
8.  }
```

♦ getResponse()

```
1.  int   PerfectPlasticMaterial::getResponse   (int   responseID,
    Information &matInfo)
2.  {
3.  Return
    PerfectPlasticMaterial::getResponse(responseID,matInfo);
4.  }
```

Step 8: Modify the "TclModelBuilderUniaxialMaterialCommand.cpp" in the "uniaxial" folder. Click [Build] to build the source code to make sure this file is modified correctly as shown in Figs. 16.13 and 16.14.

Figure 16.14 shows the following code being added to "TclModelBuilderUniaxialMaterialCommand.cpp".

```
1.  else if (strcmp(argv[1],"PerfectPlasticMaterial") == 0) {
2.  if (argc < 3) {opserr << "WARNING insufficient arguments\n";
3.  printCommand(argc,argv);
```

```
4.    opserr << "Want: uniaxialMaterial PerfectPlasticMaterial
5.    tag? " << endln;
6.    return TCL_ERROR;
7.    }
8.    int tag;double E;double Stress0;
9.    if (Tcl_GetInt(interp, argv[2], &tag) != TCL_OK) {
10.   opserr << "WARNING invalid tag in PerfectPlasticMaterial " <<
      endln;
11.   return TCL_ERROR;
12.   }
13.   if (Tcl_GetDouble(interp, argv[3], &E) != TCL_OK) {
14.   opserr << "WARNING E in PerfectPlasticMaterial " << endln;
15.   return TCL_ERROR;
16.   }
17.   if (Tcl_GetDouble(interp, argv[4], &Stress0) != TCL_OK) {
18.   opserr << "WARNING Stress0 in PerfectPlasticMaterial " <<
      endln;
19.   return TCL_ERROR;
20.   }
21.   theMaterial = new PerfectPlasticMaterial(tag,E,Stress0);
22.   }
```

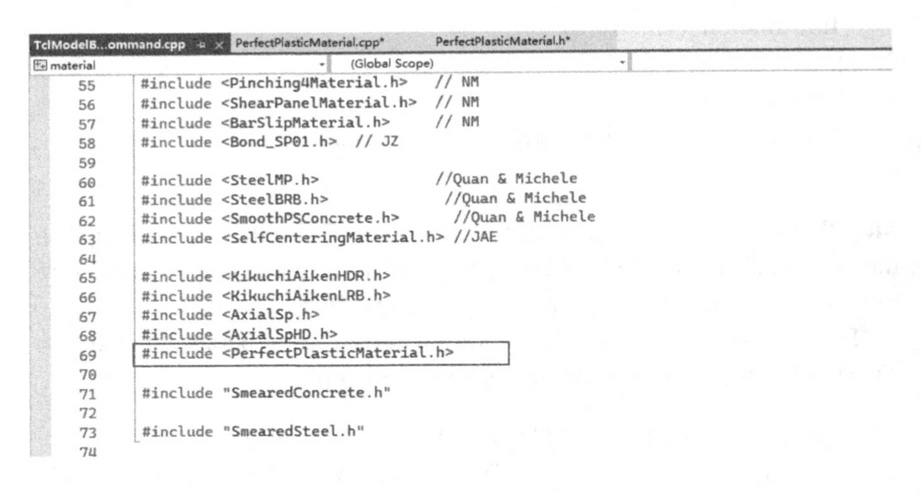

TclModelB...ommand.cpp
PerfectPlasticMaterial.cpp*
PerfectPlasticMaterial.h*
material
(Global Scope)
55 #include <Pinching4Material.h> // NM
56 #include <ShearPanelMaterial.h> // NM
57 #include <BarSlipMaterial.h> // NM
58 #include <Bond_SP01.h> // JZ
59
60 #include <SteelMP.h> //Quan & Michele
61 #include <SteelBRB.h> //Quan & Michele
62 #include <SmoothPSConcrete.h> //Quan & Michele
63 #include <SelfCenteringMaterial.h> //JAE
64
65 #include <KikuchiAikenHDR.h>
66 #include <KikuchiAikenLRB.h>
67 #include <AxialSp.h>
68 #include <AxialSpHD.h>
69 #include <PerfectPlasticMaterial.h>
70
71 #include "SmearedConcrete.h"
72
73 #include "SmearedSteel.h"
74

Fig. 16.13 Adding a header declaration.

```
TclModelB...ommand.cpp    PerfectPlasticMaterial.cpp*    PerfectPlasticMaterial.h*
material    (Global Scope)    TclModelBuilderUniaxialMaterialCommand(Clie
1109    else if (strcmp(argv[1], "PerfectPlasticMaterial") == 0) {
1110        if (argc < 3) {
1111            opserr << "WARNING insufficient arguments\n";
1112            printCommand(argc, argv);
1113            opserr << "Want: uniaxialMaterial PerfectPlasticMaterial tag? " << endln;
1114            return TCL_ERROR;
1115        }
1116        int tag; double E; double Stress0;
1117        if (Tcl_GetInt(interp, argv[2], &tag) != TCL_OK) {
1118            opserr << "WARNING invalid tag in PerfectPlasticMaterial " << endln;
1119            return TCL_ERROR;
1120        }
1121        if (Tcl_GetDouble(interp, argv[3], &E) != TCL_OK) {
1122            opserr << "WARNING E in PerfectPlasticMaterial " << endln;
1123            return TCL_ERROR;
1124        }
1125        if (Tcl_GetDouble(interp, argv[4], &Stress0) != TCL_OK) {
1126            opserr << "WARNING Stress0 in PerfectPlasticMaterial " << endln;
1127            return TCL_ERROR;
1128        }
1129        theMaterial = new PerfectPlasticMaterial(tag, E, Stress0);
1130    }
```

Fig. 16.14 Adding C++ code.

16.3 Making a Tcl Model to Test and Debug the Code

Step 1: Make a simple Tcl model and save it to a specified folder ("OpenSees\Win64\proj\opensees" in Fig. 16.15). Set breakpoints and debug the program as shown in Figs. 16.16 and 16.17.

The Tcl commands of the model are as follows:

```
1.  wipe;
2.  model basic -ndm 2 -ndf 2
3.  node 1 0.0 0.0
4.  node 2 10.0 0.0 -mass 10000.0 10000.0
5.  fix 1 1 1
6.  fix 2 0 1
7.  uniaxialMaterial PerfectPlasticMaterial 1 2e1 3.0e-2
8.  uniaxialMaterial Elastic  2  20.0
9.  element truss  1  1  2  1  1
10. element truss  2  1  2  1  2
11. recorder Node -file node2.out -time -node 2 -dof 1 2 disp
12. recorder Element -file stress1.out -time -ele 1 -material stress
13. recorder Element -file strain1.out -time -ele 1 -material strain
14. recorder Element -file stress2.out -time -ele 2 -material stress
```

Fig. 16.15 Making a test model and tabas.txt in a specified OpenSees folder.

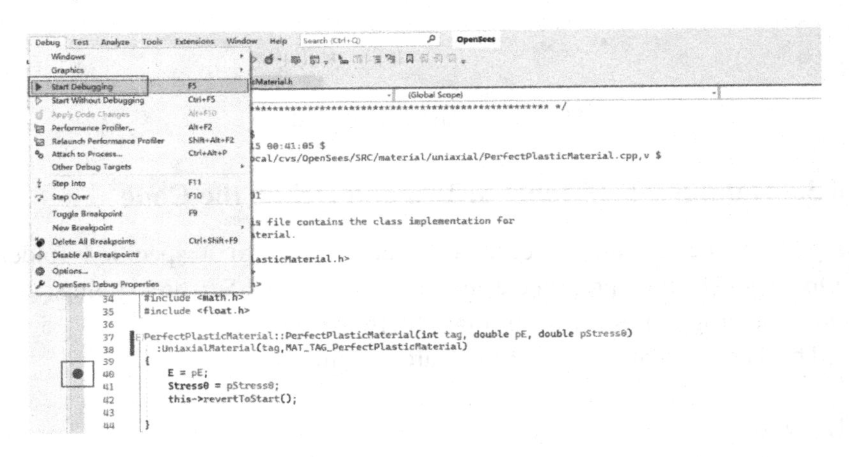

Fig. 16.16 Start debugging the program (1).

15. recorder Element -file strain2.out -time -ele 2 -material strain
16. set tabas "Path -filePath tabas.txt -dt 0.02 -factor 4"
17. pattern UniformExcitation 1 1 -accel $tabas
18. constraints Transformation
19. numberer RCM
20. test NormDispIncr 1.E-8 25 2
21. algorithm Newton
22. system BandSPD
23. integrator Newmark 0.55 0.275625
24. analysis Transient
25. analyze 1000 0.01

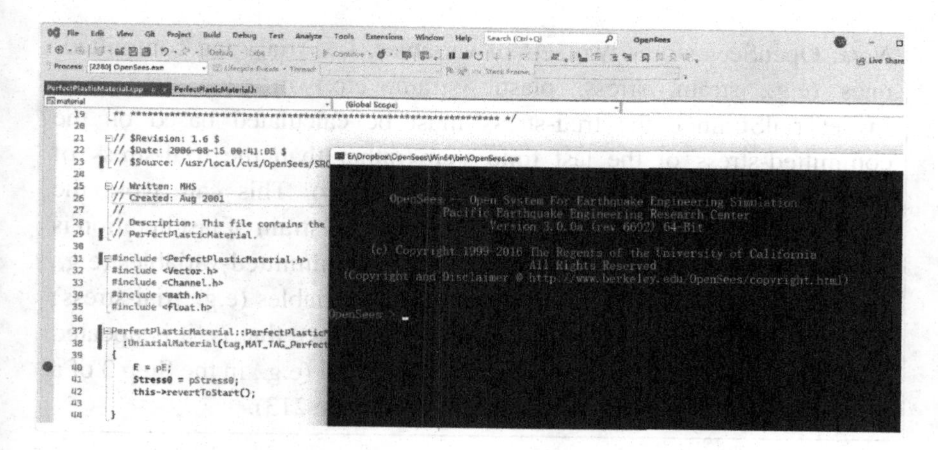

Fig. 16.17 Debugging the program (2).

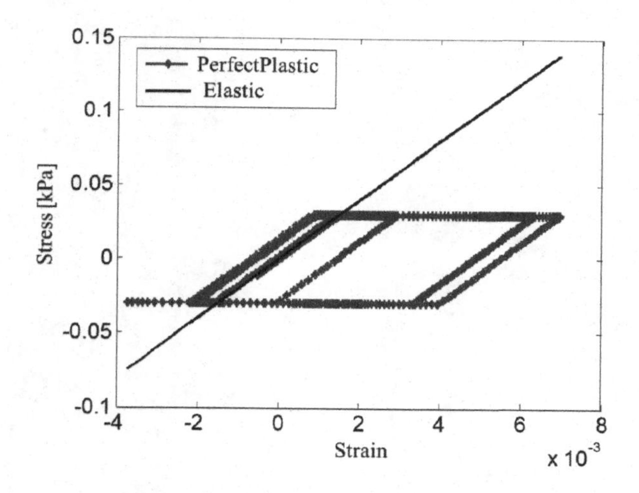

Fig. 16.18 The stress–strain responses of the newly added material.

Step 2: Use the F9 key to set breakpoints at specific lines (see Fig. 16.16). Press F5 or F10 to debug the program (see Fig. 16.17). F5 is used to start the program, and runs until the first breakpoint. F10 is used to execute the code line by line.

Step 3: The material's constitutive behavior is shown in Fig. 16.18.

Note: OpenSees stores two sets of variables, i.e., trial- and committed-ones (e.g., strain, stress, plastic strain, etc.) in each time step. In setTrialStrain(), the trial-stress must be calculated based on the committed-stress of the last time step (rather than the trial-stress of the last Newton iterations in the current step). This can avoid the possible incorrect 'unloading' case when the trial strain of last iteration is larger than trial strain of current iteration. The committed-variables (e.g., committed-stress) will be updated to the trial-variables (e.g., trial-stress) in the commitState() rather than in setTrialStrain(), that is, it is updated only after convergence of current step is achieved (e.g., in the Step 3 of a typical implicit flowchart of Section 15.3 on page 213).

Chapter 17

Adding a New Element in OpenSees

17.1 Brief Introduction

This chapter describes how to add a new element in OpenSees, i.e., an element named "corotTruss2D" (i.e., a two-dimensional (2D) corotational truss element). This element is capable of handling arbitrarily large rigid body motions with small deformation along the element [1]. The element is modified based on an existing "Truss" element in OpenSees.

The model of the corotational truss element can be found in Ref. [1] and is shown in Fig. 17.1.

In Figs. 17.1 and 17.2, Q is the element force in basic coordinate system (BCS), and the corresponding deformation in BCS is D. P_1, P_2, P_3, P_4 are the element forces in the local coordinate system (LCS), while $\bar{P}_1$, $\bar{P}_2$, $\bar{P}_3$, $\bar{P}_4$ are the element forces in the global coordinate system (GCS). U_1, U_2, U_3, U_4 are nodal displacements in LCS and $\bar{U}_1$, $\bar{U}_2$, $\bar{U}_3$, $\bar{U}_4$ are nodal displacements in GCS. L_o is the initial length of the element in reference configuration, while L_n is the length in corotated configuration. α is the angle between the local and global coordinate systems, and β is the angle of rigid body rotation. To calculate the element resisting force and tangent stiffness of the element, the following formulas are derived:

$$\bar{\mathbf{P}} = \mathbf{\Gamma}_{\mathrm{ROT}}^{T} \cdot \mathbf{\Gamma}_{\mathrm{RBM}}^{T} \cdot \mathrm{Q} \tag{17.1}$$

$$\bar{\mathbf{K}}_T = \mathbf{\Gamma}_{\mathrm{ROT}}^{T} \cdot \left(\frac{\partial \mathbf{\Gamma}_{\mathrm{RBM}}^{T}}{\partial \mathbf{U}} \cdot \mathbf{Q} + \mathbf{\Gamma}_{\mathrm{RBT}}^{T} \cdot \mathrm{k} \cdot \mathbf{\Gamma}_{\mathrm{RBM}} \right) \cdot \mathbf{\Gamma}_{\mathrm{ROT}} \tag{17.2}$$

where $\bar{\mathbf{P}}$, $\bar{\mathbf{K}}_T$ are the element resisting force, tangent stiffness in GCS, respectively. $\mathbf{\Gamma}_{\mathrm{ROT}}$ and $\mathbf{\Gamma}_{\mathrm{RBM}}$ are the rotational transformation matrix and

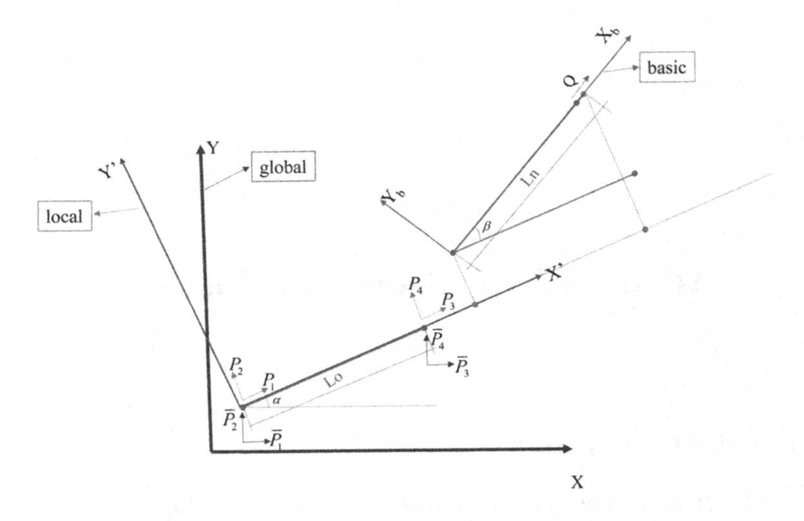

Fig. 17.1 The nodal force in basic, local, and global coordinate systems.

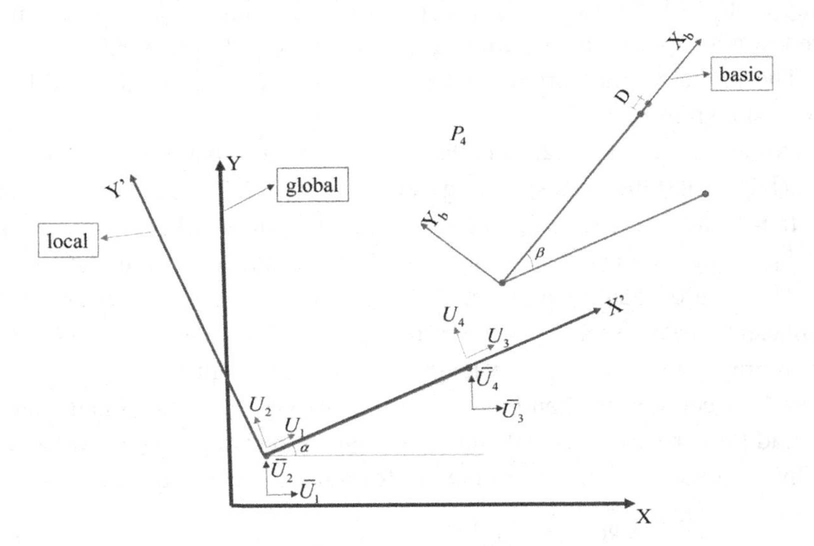

Fig. 17.2 The nodal displacements/deformation in basic, local, and global coordinate systems.

corotational transformation matrix, which are

$$\mathbf{\Gamma}_{\mathrm{ROT}} = \begin{bmatrix} \cos\alpha & \sin\alpha & 0 & 0 \\ -\sin\alpha & \cos\alpha & 0 & 0 \\ 0 & 0 & \cos\alpha & \sin\alpha \\ 0 & 0 & -\sin\alpha & \cos\alpha \end{bmatrix} \tag{17.3}$$

and

$$\mathbf{\Gamma}_{\mathrm{RBM}} = \begin{bmatrix} -\cos\beta & -\sin\beta & \cos\beta & \sin\beta \end{bmatrix} \tag{17.4}$$

Q is calculated by using the material constitutive with the deformation D of the element in BCS, and D is obtained by Eq. (17.5):

$$\mathrm{D} = (L_n - L_o)/L_o \tag{17.5}$$

$\frac{\partial \mathbf{\Gamma}^T_{\mathrm{RBM}}}{\partial \mathrm{U}} \cdot \mathrm{Q}$ is the geometric tangent stiffness. $\mathbf{\Gamma}^T_{\mathrm{RBT}} \cdot \mathrm{k} \cdot \mathbf{\Gamma}_{\mathrm{RBM}}$ is the material tangent stiffness, k is the tangent stiffness in BCS. $\frac{\partial \mathbf{\Gamma}^T_{\mathrm{RBM}}}{\partial \mathrm{U}}$ is calculated as follows:

$$\frac{\partial \mathbf{\Gamma}^T_{\mathrm{RBM}}}{\partial \mathrm{U}} = \frac{1}{L_n} \begin{bmatrix} \sin^2\beta & -\sin\beta\cdot\cos\beta & -\sin^2\beta & \sin\beta\cdot\cos\beta \\ -\sin\beta\cdot\cos\beta & \cos^2\beta & \sin\beta\cdot\cos\beta & -\cos^2\beta \\ -\sin^2\beta & \sin\beta\cdot\cos\beta & \sin^2\beta & -\sin\beta\cdot\cos\beta \\ \sin\beta\cdot\cos\beta & -\cos^2\beta & -\sin\beta\cdot\cos\beta & \cos^2\beta \end{bmatrix} \tag{17.6}$$

17.2 The Procedure for Adding the Corotational Truss Element

According to the formulas above, it is enough for us to implement the corotational truss element in OpenSees. The following steps show the procedure of adding the new element.

Step 1. Find the "Truss" element in ./OpenSees/SRC/element/truss, copy "Truss.cpp" and "Truss.h", rename them to "corotTruss2D.cpp" and "corotTruss2D.h", as shown in Fig. 17.3.

Step 2. Open Visual Studio 2017, choose view-> Solution Explorer, find element-> truss, right click on the label "truss", select Add-> Existing

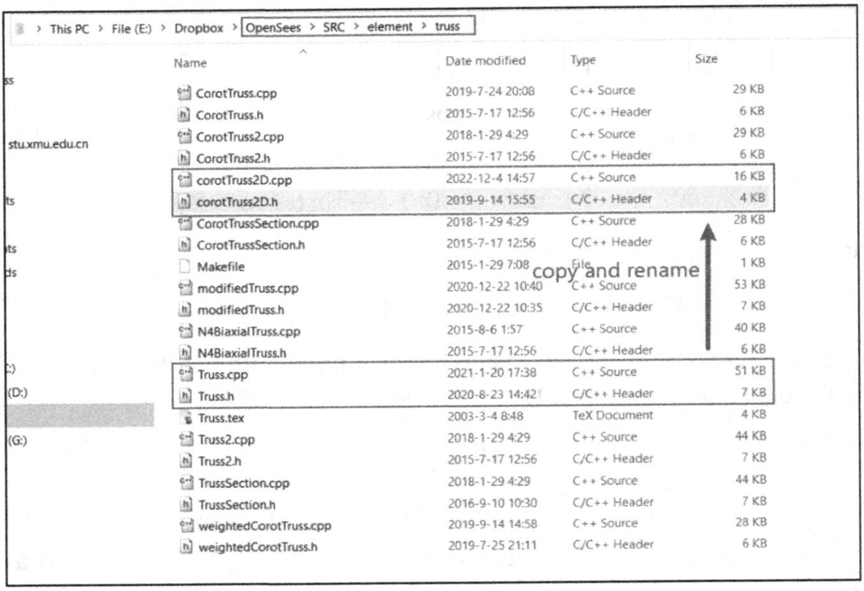

> This PC > File (E:) > Dropbox > OpenSees > SRC > element > truss

Name	Date modified	Type	Size
CorotTruss.cpp	2019-7-24 20:08	C++ Source	29 KB
CorotTruss.h	2015-7-17 12:56	C/C++ Header	6 KB
CorotTruss2.cpp	2018-1-29 4:29	C++ Source	29 KB
CorotTruss2.h	2015-7-17 12:56	C/C++ Header	6 KB
corotTruss2D.cpp	2022-12-4 14:57	C++ Source	16 KB
corotTruss2D.h	2019-9-14 15:55	C/C++ Header	4 KB
CorotTrussSection.cpp	2018-1-29 4:29	C++ Source	28 KB
CorotTrussSection.h	2015-7-17 12:56	C/C++ Header	6 KB
Makefile	2015-1-29 7:08	File	1 KB
modifiedTruss.cpp	2020-12-22 10:40	C++ Source	53 KB
modifiedTruss.h	2020-12-22 10:35	C/C++ Header	7 KB
N4BiaxialTruss.cpp	2015-8-6 1:57	C++ Source	40 KB
N4BiaxialTruss.h	2015-7-17 12:56	C/C++ Header	6 KB
Truss.cpp	2021-1-20 17:38	C++ Source	51 KB
Truss.h	2020-8-23 14:42	C/C++ Header	7 KB
Truss.tex	2003-3-4 8:48	TeX Document	4 KB
Truss2.cpp	2018-1-29 4:29	C++ Source	44 KB
Truss2.h	2015-7-17 12:56	C/C++ Header	7 KB
TrussSection.cpp	2018-1-29 4:29	C++ Source	44 KB
TrussSection.h	2016-9-10 10:30	C/C++ Header	7 KB
weightedCorotTruss.cpp	2019-9-14 14:58	C++ Source	28 KB
weightedCorotTruss.h	2019-7-25 21:11	C/C++ Header	6 KB

copy and rename

Fig. 17.3 Copying and renaming the files.

Item, choose "corotTruss2D.cpp" and "corotTruss2D.h" and add them into the element project, as shown in Figs. 17.4 and 17.5.

Step 3. Replace the string "Truss" by "corotTruss2D" in "corotTruss2D.cpp" and "corotTruss2D.h", as shown in Fig. 17.6.

Step 4. Add an element tag in OpenSees. To find the file "classTags.h", the easiest way is opening a .cpp file in the project of element (e.g., "Truss.cpp" as shown in Fig. 17.7), right click on the string "ELE_TAG_Truss" and choose "Go to Definition". Add the code "#define ELE_TAG_corotTruss2D 502" in "classTags.h" as shown in Fig. 17.8. The number "502 " should be a unique number, different from other element tags.

Step 5. Find "TclElementCommands.cpp" in the element project (i.e., element-> Source Files, see Fig. 17.9) to build the interface of the element. "*extern void *OPS_weightedCorotTrussElement(void);*" is added in

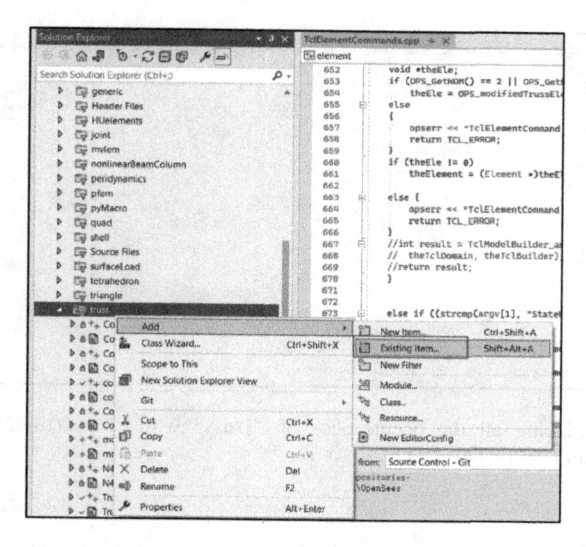

Fig. 17.4 Adding "corotTruss2D.cpp" and "corotTruss2D.h" into "Solution Explorer" (1).

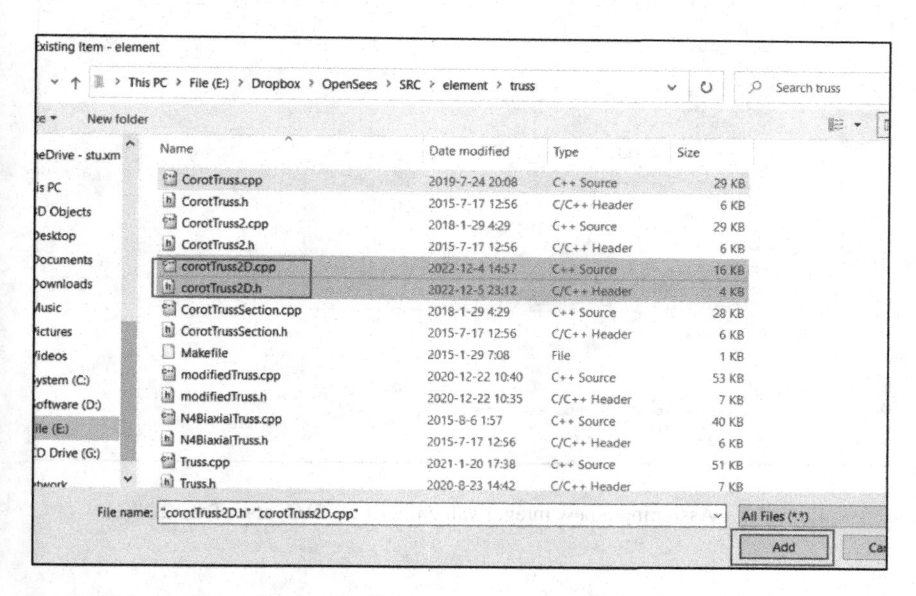

Fig. 17.5 Adding "corotTruss2D.cpp" and "corotTruss2D.h" into "Solution Explorer" (2).

```
corotTruss2D.h    corotTruss2D.cpp
element    (Global Scope)
10  ** University of California, Berkeley, is strictly prohibited.  See   **
11  ** file 'COPYRIGHT'  in main directory for information on usage and   **
12  ** redistribution,  and for a DISCLAIMER OF ALL WARRANTIES.            **
13  **                                                                    **
14  ** Developed by:                                                      **
15  **   Frank McKenna (fmckenna@ce.berkeley.edu)                         **
16  **   Gregory L. Fenves (fenves@ce.berkeley.edu)                       **
17  **   Filip C. Filippou (filippou@ce.berkeley.edu)                     **
18  **                                                                    **
19  ** ****************************************************************** */
20
21  // $Revision$
22  // $Date$
23  // $URL$
24
25
26  #ifndef Truss_h
27  #define Truss_h
28
29
30  #include <Element.h>
31  #include <Matrix.h>
32
33  class Node;
34  class Channel;
35  class UniaxialMaterial;
36
37  class Truss : public Element
38  {
```

```
Truss
corotTruss2D
All open documents
```

Fig. 17.6 Replacing all the occurrences of "Truss" with "corotTruss2D" in "corotTruss2D.cpp" and "corotTruss2D.h".

```
Truss.cpp    corotTruss2D.h    corotTruss2D.cpp
element    Truss    Truss(int
160         " $iNode $jNode $A $matTag <-rho $rho> <-cMass $flag> <-doRayleigh $flag>\n";
161     }
162
163     return theElement;
164 }
165
166 Truss::Truss(int tag, int dim,
167              int Nd1, int Nd2,
168              UniaxialMaterial &theMat,
169              double a, double r,
170              int damp, int cm)
171  :Element(tag, ELE_TAG_Truss),
172   theMaterial(0), con
173   dimension(dim), num
174   theLoad(0), theMatr
175   L(0.0), A(a), rho(r
176   cMass(cm), initialD
177 {
178     // get a copy of
179     theMaterial = the
180     if (theMaterial =
181       opserr << "FATA
182     "failed to get a                                            endln;
183       exit(-1);
184     }
185
186     // ensure the con                                           ues
187     if (connectedExte
188       opserr << "FATA                                           ID of size 2\n";
189       exit(-1);
```

```
Quick Actions and Refactorings...   Ctrl+.
Rename...                           Ctrl+R, Ctrl+R
Generate Graph of Include Files
View Code                           F7
Peek Definition                     Alt+F12
Go To Definition                    F12
Go To Declaration                   Ctrl+F12
Find All References                 Shift+F12
View Call Hierarchy                 Ctrl+K, Ctrl+T
Peek Header / Code File             Ctrl+K, Ctrl+J
Toggle Header / Code File           Ctrl+K, Ctrl+O
Breakpoint
```

Fig. 17.7 Assigning a new integer value to "ELE_TAG_corotTruss2D"(1).

```
classTags.h   Truss.cpp   corotTruss2D.h   corotTruss2D.cpp
element                                   (Global Scope)
692   #define ELE_TAG_MultipleNormalSpring     159
693   #define ELE_TAG_KikuchiBearing           160
694   #define ELE_TAG_YamamotoBiaxialHDR       161
695   #define ELE_TAG_MVLEM                    162
696   #define ELE_TAG_SFI_MVLEM                163
697   #define ELE_TAG_PFEMElement2DFIC         164
698   #define ELE_TAG_ElastomericBearingBoucWenMod3d 165
699   #define ELE_TAG_FPBearingPTV             166
700   #define ELE_TAG_ShellDKGT                167
701   #define ELE_TAG_ShellNLDKGT              168
702   #define ELE_TAG_CatenaryCable            169
703   #define ELE_TAG_DispBeamColumn3dThermal  170  // L.Jiang [SIF]
704   #define ELE_TAG_ForceBeamColumn2dThermal 171  //L.Jiang [SIF]
705   #define ELE_TAG_ForceBeamColumn3dThermal 172  //L.Jiang [SIF] //Still testing
706   #define ELE_TAG_ShellMITC4Thermal        173   //L.Jiang [SIF]
707   #define ELE_TAG_ShellNLDKGQThermal       174   //L.Jiang [SIF]
708   #define ELE_TAG_ShellANDeS               175   //L.Jiang [SIF]
709   #define ELE_TAG_AxEqDispBeamColumn2d     178
710   #define ELE_TAG_FourNodeTetrahedron      179
711   #define ELE_TAG_TriSurfaceLoad           180
712   #define ELE_TAG_QuadBeamEmbedContact     500
713   #define ELE_TAG_bondbasedPeridynamic     501  //Lei
714   #define ELE_TAG_corotTruss2D         502 //Lei
715   #define ELE_TAG_stateBasedPeridynamic 503  //Lei
716   #define ELE_TAG_modifiedTruss 504  //Lei
717
```

Fig. 17.8 Assigning a new integer value to "ELE_TAG_corotTruss2D"(2).

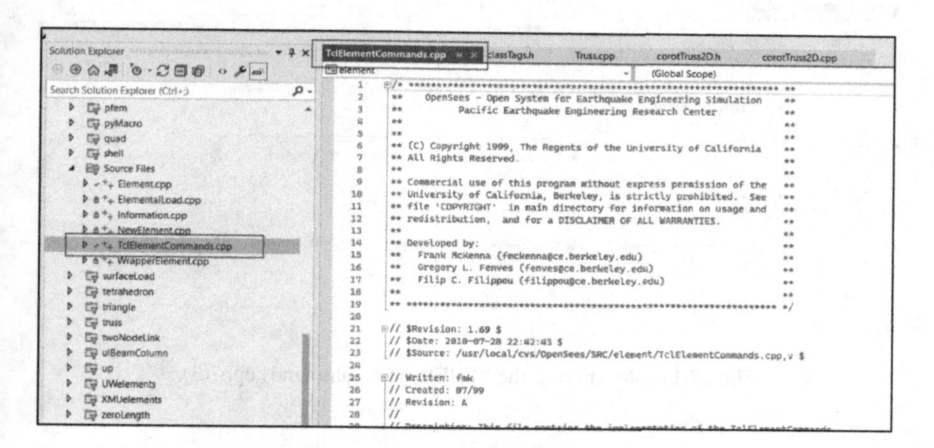

Fig. 17.9 Modifying the "TclElementCommands.cpp"(1).

the corresponding places as shown in Fig. 17.10. Besides, the following codes are also needed as shown in Fig. 17.11:

1. else if ((strcmp(argv[1], "corotTruss2D") == 0) || strcmp(argv[1], "corotTruss2D") == 0) {
2. void *theEle = OPS_corotTruss2DElement();
3. if (theEle != 0)

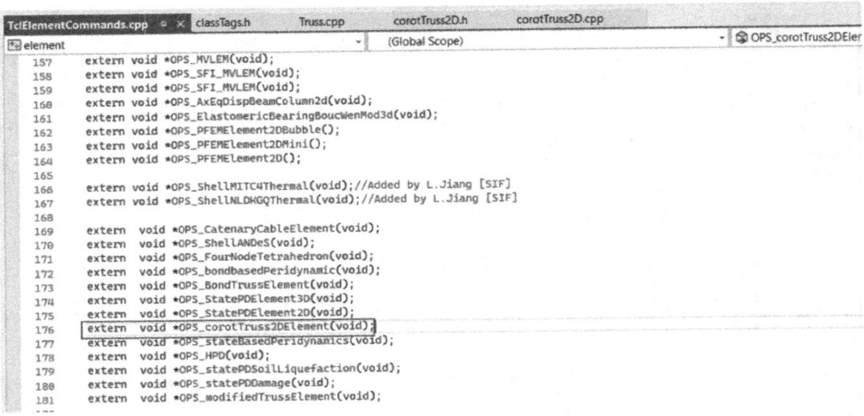

```
TclElementCommands.cpp   classTags.h   Truss.cpp   corotTruss2D.h   corotTruss2D.cpp
element        (Global Scope)        OPS_corotTruss2DEler
157  extern void *OPS_MVLEM(void);
158  extern void *OPS_SFI_MVLEM(void);
159  extern void *OPS_SFI_MVLEM(void);
160  extern void *OPS_AxEqDispBeamColumn2d(void);
161  extern void *OPS_ElastomericBearingBoucWenMod3d(void);
162  extern void *OPS_PFEMElement2DBubble();
163  extern void *OPS_PFEMElement2DMini();
164  extern void *OPS_PFEMElement2D();
165
166  extern void *OPS_ShellMITC4Thermal(void);//Added by L.Jiang [SIF]
167  extern void *OPS_ShellNLDKGQThermal(void);//Added by L.Jiang [SIF]
168
169  extern  void *OPS_CatenaryCableElement(void);
170  extern  void *OPS_ShellANDeS(void);
171  extern  void *OPS_FourNodeTetrahedron(void);
172  extern  void *OPS_bondbasedPeridynamic(void);
173  extern  void *OPS_BondTrussElement(void);
174  extern  void *OPS_StatePDElement3D(void);
175  extern  void *OPS_StatePDElement2D(void);
176  extern  void *OPS_corotTruss2DElement(void);
177  extern  void *OPS_stateBasedPeridynamics(void);
178  extern  void *OPS_HPD(void);
179  extern  void *OPS_statePDSoilLiquefaction(void);
180  extern  void *OPS_statePDDamage(void);
181  extern  void *OPS_modifiedTrussElement(void);
```

Fig. 17.10 Modifying the “TclElementCommands.cpp”(2).

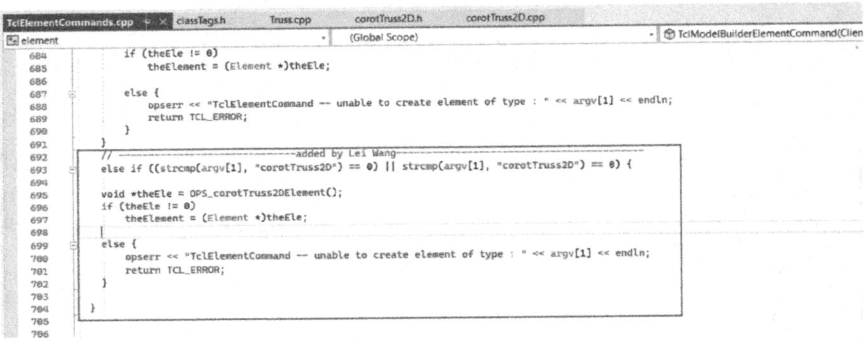

```
TclElementCommands.cpp   classTags.h   Truss.cpp   corotTruss2D.h   corotTruss2D.cpp
element        (Global Scope)        TclModelBuilderElementCommand(Clien
684          if (theEle != 0)
685              theElement = (Element *)theEle;
686
687          else {
688              opserr << "TclElementCommand -- unable to create element of type : " << argv[1] << endln;
689              return TCL_ERROR;
690          }
691      }
692      //------------------------added by Lei Wang--------------------------
693      else if ((strcmp(argv[1], "corotTruss2D") == 0) || strcmp(argv[1], "corotTruss2D") == 0) {
694
695      void *theEle = OPS_corotTruss2DElement();
696      if (theEle != 0)
697          theElement = (Element *)theEle;
698
699      else {
700          opserr << "TclElementCommand -- unable to create element of type : " << argv[1] << endln;
701          return TCL_ERROR;
702      }
703
704   }
705
706
```

Fig. 17.11 Modifying the “TclElementCommands.cpp”(3).

```
4. theElement = (Element *)theEle;
5. else {
6. opserr << "TclElementCommand -- unable to create element
   of type : " << argv[1] << endln;
7. return TCL_ERROR;
8. }
9. }
```

```
25
26  #ifndef corotTruss2D_h
27  #define corotTruss2D_h
28
29
30  #include <Element.h>
31  #include <Matrix.h>
32
33  class Node;
34  class Channel;
35  class UniaxialMaterial;
36
37  class corotTruss2D : public Element
38  {
39  public:
```

```
90 %    No issues found
Output
Show output from: Build
6>Project not selected to build for this solution configuration
7>------ Skipped Build: Project: OpenSeesTk, Configuration: Release x64 ------
7>Project not selected to build for this solution configuration
8>------ Build started: Project: OpenSees, Configuration: Release x64 ------
8>   Creating library .\..\..\bin\OpenSees.lib and object .\..\..\bin\OpenSees.exp
8>OpenSees.vcxproj -> F:\Dropbox\OpenSees\Win64\bin\OpenSees.exe
========== Build: 2 succeeded, 0 failed, 23 up-to-date, 6 skipped ==========
```

Fig. 17.12 Building the solution until it is built successfully.

Step 6. Build the solution to see if there are any errors. If so, correct them until the solution is built successfully as shown in Fig. 17.12, then it is possible to add/modify the member functions in the new element.

Step 7. Modify the "corotTruss2D.cpp" and "corotTruss2D.h" files to implement the new element. It is easier to do the job by modifying the Truss element. In the file "corotTruss2D.h", it is necessary to add/delete some variables/member functions. The codes of "corotTruss2D.h" are as follows:

```
1.  #ifndef corotTruss2D_h
2.  #define corotTruss2D_h
3.  #include <Element.h>
4.  #include <Matrix.h>
5.  class Node;
6.  class Channel;
7.  class UniaxialMaterial;
8.  class corotTruss2D : public Element
9.  {
10. public:
```

```
11. corotTruss2D(int tag, int dimension,
12. int Nd1, int Nd2,
13. UniaxialMaterial &theMaterial,
14. double A);
15. corotTruss2D();
16. ~corotTruss2D();
17. const char *getClassType(void) const { return "corotTruss2D"; };
18. int getNumExternalNodes(void) const;
19. const ID &getExternalNodes(void);
20. Node **getNodePtrs(void);
21. int getNumDOF(void);
22. void setDomain(Domain *theDomain);
23. int commitState(void);
24. int revertToLastCommit(void);
25. int revertToStart(void);
26. int update(void);
27. const Matrix &getDamp(void);
28. const Matrix &getMass(void);
29. const Matrix &getInitialStiff(void);
30. const Matrix &getTangentStiff(void);
31. const Vector &getResistingForce(void);
32. const Matrix &getRBM(void);
33. const Matrix &getRBMTpartialU(void);
34. const Matrix &getR(void);
35. int sendSelf(int commitTag, Channel &theChannel);
36. int recvSelf(int commitTag, Channel &theChannel,
    FEM_ObjectBroker &theBroker);
37. int displaySelf(Renderer &, int mode, float fact, const char
    **displayModes = 0, int numModes = 0);
38. void Print(OPS_Stream &s, int flag = 0);
39. Response *setResponse(const char **argv, int argc,
    OPS_Stream &s);
40. int getResponse(int responseID, Information &eleInformation);
41. protected:
42. private:
43. UniaxialMaterial *theMaterial; // pointer to a material
```

```
44. ID connectedExternalNodes;  // contains the tags of the end
    nodes
45. int dimension;          // corotTruss2D in 2 or 3d domain
46. int numDOF;                // number of dof for corotTruss2D
47. Matrix *theMatrix;
48. Vector *theVector;
49. double Lo;      // length based on undeformed configuration
50. double Ln;
51. double A;        //  area of corotTruss2D
52. double cosX[3]; //  direction cosines
53. Node *theNodes[2];
54. double cosBeta;
55. double sinBeta;
56. static Matrix corotTruss2DM4; // class wide matrix for 4*4
57. static Vector corotTruss2DV4; // class wide Vector for size 4
58. };
59. #endif
```

In this example, the tcl command for the new element is designed as "*element corotTruss2D $tag $iNode $jNode $A $matTag*". The codes for interpreting the tcl command are inside a global function *OPS_corotTruss2DElement ()*:

```
1. OPS_Export void *
2. OPS_corotTruss2D()
3. {
4. Element *theElement = 0;
5. int numRemainingArgs = OPS_GetNumRemainingInputArgs();
6. if (numRemainingArgs < 4) {
7. opserr << "Invalid Args want: element corotTruss2D $tag $iNode
   $jNode $sectTag \n";
8. opserr << " or: element corotTruss2D $tag $iNode $jNode $A
   $matTag\n";
9. return 0;
10. }
11. int iData[3];
12. double A = 0.0;
```

```
13. int matTag = 0;
14. int ndm = OPS_GetNDM();
15. int numData = 3;
16. if (OPS_GetInt(&numData, iData) != 0) {
17. opserr << "WARNING invalid integer (tag, iNode, jNode) in
     element corotTruss2D " >> endln;
18. return 0;
19. }
20. numData = 1;
21. if (OPS_GetDouble(&numData, &A) != 0) {
22. opserr << "WARNING: Invalid A: element corotTruss2D " <<
    iData[0] <<
23. "$iNode $jNode $A $matTag\n";
24. return 0;
25. }
26. numData = 1;
27. if (OPS_GetInt(&numData, &matTag) != 0) {
28. opserr << "WARNING: Invalid matTag: element corotTruss2D "
     << iData[0] <<
29. " $iNode $jNode $A $matTag\n";
30. return 0;
31. }
32. UniaxialMaterial *theUniaxialMaterial =
     OPS_GetUniaxialMaterial(matTag);
33. if (theUniaxialMaterial == 0) {
34. opserr << "WARNING: Invalid material not found element
    corotTruss2D " << iData[0] << " $iNode $jNode $A " << matTag
     << "\n";
35. return 0;
36. }
37. theElement = new corotTruss2D(iData[0], ndm, iData[1],
    iData[2], *theUniaxialMaterial, A);
38. if (theElement == 0) {
39. opserr << "WARNING: out of memory: element corotTruss2D "
    <<iData[0] <<
```

```
40.  " $iNode $jNode $A $matTag\n";
41.  }
42.  return theElement;
43. }
```

The constructor function *corotTruss2D ()* is called when the object is created, before any other functions are called. The constructor function initializes the function as follows:

```
1.  corotTruss2D::corotTruss2D(int tag, int dim,
2.  int Nd1, int Nd2,
3.  UniaxialMaterial &theMat,
4.  double a)
5.  :Element(tag,ELE_TAG_corotTruss2D),
6.  theMaterial(0), connectedExternalNodes(2),
7.  dimension(dim), numDOF(0),A(a),
8.  theMatrix(0), theVector(0), Lo(0.0)
9.  {
10. theMaterial = theMat.getCopy();
11. if (theMaterial == 0) {
12. opserr << "FATAL corotTruss2D::corotTruss2D - " << tag <<
13. "failed to get a copy of material with tag " << theMat.getTag() <<
    endln;
14. exit(-1);
15. }
16. if (connectedExternalNodes.Size() != 2) {
17. opserr << "FATAL corotTruss2D::corotTruss2D - " << tag <<
    "failed to create an ID of size 2\n";
18. exit(-1);
19. }
20. connectedExternalNodes(0) = Nd1;
21. connectedExternalNodes(1) = Nd2;
22. for (int i=0; i < 2; i++)
23. theNodes[i] = 0;
24. cosX[0] = 0.0;
25. cosX[1] = 0.0;
26. cosX[2] = 0.0;
27. cosBeta = 0.0;
28. sinBeta = 0.0;
29. }
```

The function "*void corotTruss2D::setDomain(Domain *theDomain)*" is used to add the element to the domain by calling *the base class method, i.e., this-> DomainComponent::setDomain(theDomain)*. This function will also get the pointer of each node of this element from the domain and store them inside the element. Then some initialization work can be done here (e.g., calculate the initial length and angle of this truss):

```
1.  void
2.  corotTruss2D::setDomain(Domain *theDomain)
3.  {
4.  if (theDomain == 0) {
5.  theNodes[0] = 0;
6.  theNodes[1] = 0;
7.  Lo = 0;
8.  return;
9.  }
10. int Nd1 = connectedExternalNodes(0);
11. int Nd2 = connectedExternalNodes(1);
12. theNodes[0] = theDomain-> getNode(Nd1);
13. theNodes[1] = theDomain-> getNode(Nd2);
14. if ((theNodes[0] == 0) || (theNodes[1] == 0)) {
15. if (theNodes[0] == 0)
16. opserr << "corotTruss2D::setDomain() - corotTruss2D" << this-
    >getTag() << " node " << Nd1 << "does not exist in the model\n";
17. else
18. opserr << "corotTruss2D::setDomain() - corotTruss2D" << this-
    >getTag() << " node " << Nd2 << "does not exist in the
    model\n";
19. return;
20. }
21. this-> DomainComponent::setDomain(theDomain);
22. numDOF = 4;
23. theMatrix = &corotTruss2DM4;
24. theVector = &corotTruss2DV4;
25. const Vector &end1Crd = theNodes[0]-> getCrds();
26. const Vector &end2Crd = theNodes[1]-> getCrds();
```

```
27. const Vector &end1Disp = theNodes[0]->getDisp();
28. const Vector &end2Disp = theNodes[1]->getDisp();
29. double dx = end2Crd(0)-end1Crd(0);
30. double dy = end2Crd(1)-end1Crd(1);
31. double iDispX = end2Disp(0)-end1Disp(0);
32. double iDispY = end2Disp(1)-end1Disp(1);
33. Lo = sqrt(dx*dx + dy*dy);
34. Ln = Lo;
35. if (Lo == 0.0) {
36. opserr<<"WARNING corotTruss2D::setDomain() -
    corotTruss2D "<< this->getTag()<<" has zero length\n";
37. return;
38. }
39. cosX[0] = dx/Lo;
40. cosX[1] = dy/Lo;
41. }
```

The function "*int corotTruss2D::update(void)*" is called after the nodal displacements are updated. For example, if Newton's method is employed, a new trial displacement is calculated and *update(void)* will be called in each iteration (The system may not converge in this iteration). In element level, the update() will calculate the trial strain (and perhaps also the trial strain rate if needed) at each material point, and call the material to set the trail strain (and strain rate, i.e., *setTrialStrain(strain, rate)* or *setTrialStrain(strain)*). The cosine and sine of the rigid body rotation are also calculated here.

```
1. int
2. corotTruss2D::update(void)
3. {
4. const Vector &end1Disp = theNodes[0]->getTrialDisp();
5. const Vector &end2Disp = theNodes[1]->getTrialDisp();
6. const Vector &crdNode1 = theNodes[0]->getCrds();
7. const Vector &crdNode2 = theNodes[1]->getCrds();
8. Vector currentCrd1 = end1Disp + crdNode1;
9. Vector currentCrd2 = end2Disp + crdNode2;
```

```
10. Ln = sqrt(pow(currentCrd1(0) - currentCrd2(0), 2) +
    pow(currentCrd1(1) - currentCrd2(1), 2));
11. double strain = (Ln - Lo) / Lo;
12. double cosab = (currentCrd2(0) - currentCrd1(0)) / Ln;
13. double sinab = (currentCrd2(1) - currentCrd1(1)) / Ln;
14. Vector cossinab(2);
15. cossinab(0) = cosab;
16. cossinab(1) = sinab;
17. Matrix trans(2, 2);
18. trans.Zero();
19. trans(0, 0) = cosX[0];
20. trans(0, 1) = -cosX[1];
21. trans(1, 0) = cosX[1];
22. trans(1, 1) = cosX[0];
23. Matrix transInv(2, 2);
24. trans.Invert(transInv);
25. Vector cosb(2);
26. cosb.Zero();
27. cosb.addMatrixVector(1.0, transInv, cossinab,1.0);
28. cosBeta = cosb(0);
29. sinBeta = cosb(1);
30. return theMaterial->setTrialStrain(strain);
31. }
```

The function "*const Matrix &corotTruss2D::getTangentStiff(void)*" is used to obtain the global tangent stiffness of the element, which is calculated using Eq. (17.2).

```
1. const Matrix &
2. corotTruss2D::getTangentStiff(void)
3. {
4. Matrix &kg = *theMatrix;
5. kg.Zero();
6. static Matrix kgm(4, 4);
7. static Matrix kgt(4, 4);
8. kgt.Zero();
9. kgm.Zero();
```

```
10.  double EA = A * theMaterial->getTangent();
11.  EA /= Lo ;
12.  Matrix RBM = this->getRBM();
13.  kgm.addMatrixTransposeProduct(1.0, RBM, RBM, EA);
    //kgm=RBM'*EA*RBM
14.  double q = A * theMaterial->getStress();
15.  double SL = q / Ln;
16.  Matrix RBMpartialU = this->getRBMTpartialU();
17.  kgt.addMatrix(0.0, RBMpartialU, SL);
18.  kg.addMatrix(0.0, kgm, 1.0);
19.  kg.addMatrix(1.0, kgt, 1.0);
20.  Matrix R = this->getR();
21.  kg.addMatrixTripleProduct(0.0, R, kg, 1.0);    //R'*kg*R*1.0
22.  return *theMatrix;
23.  }
```

The functions *"getRBM()"*, *"getRBMTpartialU()"*, *"getR()"* are defined as follows:

```
1.  const Matrix &
2.  corotTruss2D::getRBM(void)
3.  {
4.  static Matrix rbm(1,4);
5.  rbm.Zero();
6.  rbm(0,0) = -cosBeta;
7.  rbm(0,1) = -sinBeta;
8.  rbm(0,2) = cosBeta;
9.  rbm(0,3) = sinBeta;
10. return rbm;
11. }
```

```
1. const Matrix &
2. corotTruss2D::getRBMTpartialU(void)
3. {
4. static Matrix rbmm (4,4);
5. rbmm.Zero();
6. double s2 = pow(sinBeta, 2);
```

```
7.  double sc = sinBeta * cosBeta;
8.  double c2 = pow(cosBeta, 2);
9.  rbmm(0, 0) = s2; rbmm(0, 1) = -sc ; rbmm(0, 2) = -s2 ;
    rbmm(0, 3) = sc ;
10. rbmm(1, 0) = -sc ; rbmm(1, 1) = c2 ;rbmm(1, 2) = sc ;
    rbmm(1, 3) = -c2 ;
11. rbmm(2, 0) = -s2 ; rbmm(2, 1) = sc ;rbmm(2, 2) = s2 ;
    rbmm(2, 3) = -sc ;
12. rbmm(3, 0) = sc ; rbmm(3, 1) = -c2 ;rbmm(3, 2) = -sc ;
    rbmm(3, 3) = c2 ;
13. return rbmm;
14. }
```

```
1.  const Matrix &
2.  corotTruss2D::getR(void)
3.  {
4.  static Matrix r(4, 4);
5.  r.Zero();
6.  double cosx = cosX[0];
7.  double sinx = cosX[1];
8.  r(0, 0) = cosx; r(1, 0) = -sinx;
9.  r(0, 1) = sinx; r(1, 1) = cosx;
10. r(2, 2) = cosx; r(3, 2) = -sinx;
11. r(2, 3) = sinx; r(3, 3) = cosx;
12. return r;
13. }
```

The function *"const Vector &corotTruss2D::getResistingForce()"* is used to obtain the resisting force at element level, which can be calculated using Eq (17.1).

```
1.  const Vector &
2.  corotTruss2D::getResistingForce()
3.  {
4.  double force = A*theMaterial->getStress(); //basic Q
5.  Vector transRBM(4);
6.  transRBM(0) = -cosBeta;
```

```
7.    transRBM(1) = -sinBeta;
8.    transRBM(2) = cosBeta;
9.    transRBM(3) = sinBeta;
10.   Matrix R = this-> getR();
11.   Vector &F=*theVector;
12.   F.Zero();
13.   F.addMatrixTransposeVector(0.0, R, transRBM, force);
      //R'*transRBM*force
14.   return *theVector;
15.   }
```

It is worth mentioning that all functions in the *class corotTruss2D* should be implemented. However, since this chapter is aimed to show how to add an element in OpenSees, only the most important functions are introduced herein, e.g., *setDomain()*, *update(void)*, *getTangentStiff(void)*, and *getResistingForce()*, etc. For the other functions that are relatively simple, e.g., *getMass(void)*, *zeroLoad(void)*, users can implement them by referring to the existing elements in OpenSees. For the functions that are simple but not used in the example (e.g., static analysis), such as *getInitialStiff(void)* and *getDamp(void)*, users can derive the formulas and implement them in the corresponding functions if needed later. Currently, only a dummy function is implemented for each of them, e.g.,

```
1.    const Matrix &
2.    weightedCorotTruss::getDamp(void)
3.    {
4.    opserr<<''warning: damping is not implemented in the
      current version!''<<endln;
5.    Matrix &K = *theMatrix;
6.    K.Zero();
7.    return *theMatrix;
8.    }
```

17.3 Numerical Examples

After successfully implementing the element, it is necessary to build a model to verify the element. The geography of the model is shown

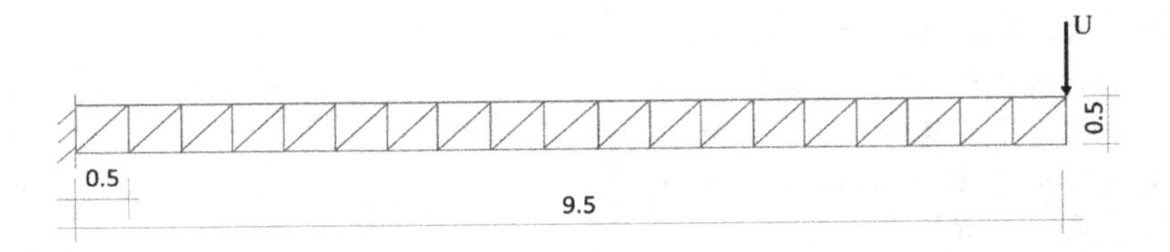

Fig. 17.13 The geography of the model.

in Fig. 17.13. There are 40 nodes and 77 elements in the simple model. The tcl commands are as follows.

```
1.  wipe
2.  model basic -ndm 2 -ndf 2
3.  node  1     0     0
4.  node  2     0.5   0
5.  node  3     1     0
6.  node  4     1.5   0
7.  node  5     2     0
8.  node  6     2.5   0
9.  node  7     3     0
10. node  8     3.5   0
11. node  9     4     0
12. node  10    4.5   0
13. node  11    5     0
14. node  12    5.5   0
15. node  13    6     0
16. node  14    6.5   0
17. node  15    7     0
18. node  16    7.5   0
19. node  17    8     0
20. node  18    8.5   0
21. node  19    9     0
22. node  20    9.5   0
23. node  21    0     0.5
24. node  22    0.5   0.5
25. node  23    1     0.5
26. node  24    1.5   0.5
```

```
27.  node  25    2    0.5
28.  node  26    2.5  0.5
29.  node  27    3    0.5
30.  node  28    3.5  0.5
31.  node  29    4    0.5
32.  node  30    4.5  0.5
33.  node  31    5    0.5
34.  node  32    5.5  0.5
35.  node  33    6    0.5
36.  node  34    6.5  0.5
37.  node  35    7    0.5
38.  node  36    7.5  0.5
39.  node  37    8    0.5
40.  node  38    8.5  0.5
41.  node  39    9    0.5
42.  node  40    9.5  0.5
43.  fix 1 1 1
44.  fix 21 1 1
45.  uniaxialMaterial Elastic 1 20
46.  element corotTruss2D 1     1   2   1  1
47.  element corotTruss2D 2     2   3   1  1
48.  element corotTruss2D 3     3   4   1  1
49.  element corotTruss2D 4     4   5   1  1
50.  element corotTruss2D 5     5   6   1  1
51.  element corotTruss2D 6     6   7   1  1
52.  element corotTruss2D 7     7   8   1  1
53.  element corotTruss2D 8     8   9   1  1
54.  element corotTruss2D 9     9   10  1  1
55.  element corotTruss2D 10    10  11  1  1
56.  element corotTruss2D 11    11  12  1  1
57.  element corotTruss2D 12    12  13  1  1
58.  element corotTruss2D 13    13  14  1  1
59.  element corotTruss2D 14    14  15  1  1
60.  element corotTruss2D 15    15  16  1  1
61.  element corotTruss2D 16    16  17  1  1
62.  element corotTruss2D 17    17  18  1  1
```

```
63.  element corotTruss2D 18   18  19  1  1
64.  element corotTruss2D 19   19  20  1  1
65.  element corotTruss2D 20   21  22  1  1
66.  element corotTruss2D 21   22  23  1  1
67.  element corotTruss2D 22   23  24  1  1
68.  element corotTruss2D 23   24  25  1  1
69.  element corotTruss2D 24   25  26  1  1
70.  element corotTruss2D 25   26  27  1  1
71.  element corotTruss2D 26   27  28  1  1
72.  element corotTruss2D 27   28  29  1  1
73.  element corotTruss2D 28   29  30  1  1
74.  element corotTruss2D 29   30  31  1  1
75.  element corotTruss2D 30   31  32  1  1
76.  element corotTruss2D 31   32  33  1  1
77.  element corotTruss2D 32   33  34  1  1
78.  element corotTruss2D 33   34  35  1  1
79.  element corotTruss2D 34   35  36  1  1
80.  element corotTruss2D 35   36  37  1  1
81.  element corotTruss2D 36   37  38  1  1
82.  element corotTruss2D 37   38  39  1  1
83.  element corotTruss2D 38   39  40  1  1
84.  element corotTruss2D 39   1   21  1  1
85.  element corotTruss2D 40   2   22  1  1
86.  element corotTruss2D 41   3   23  1  1
87.  element corotTruss2D 42   4   24  1  1
88.  element corotTruss2D 43   5   25  1  1
89.  element corotTruss2D 44   6   26  1  1
90.  element corotTruss2D 45   7   27  1  1
91.  element corotTruss2D 46   8   28  1  1
92.  element corotTruss2D 47   9   29  1  1
93.  element corotTruss2D 48   10  30  1  1
94.  element corotTruss2D 49   11  31  1  1
95.  element corotTruss2D 50   12  32  1  1
96.  element corotTruss2D 51   13  33  1  1
97.  element corotTruss2D 52   14  34  1  1
98.  element corotTruss2D 53   15  35  1  1
```

```
99.  element corotTruss2D 54    16  36  1  1
100. element corotTruss2D 55    17  37  1  1
101. element corotTruss2D 56    18  38  1  1
102. element corotTruss2D 57    19  39  1  1
103. element corotTruss2D 58    20  40  1  1
104. element corotTruss2D 59    1   22  1  1
105. element corotTruss2D 60    2   23  1  1
106. element corotTruss2D 61    3   24  1  1
107. element corotTruss2D 62    4   25  1  1
108. element corotTruss2D 63    5   26  1  1
109. element corotTruss2D 64    6   27  1  1
110. element corotTruss2D 65    7   28  1  1
111. element corotTruss2D 66    8   29  1  1
112. element corotTruss2D 67    9   30  1  1
113. element corotTruss2D 68    10  31  1  1
114. element corotTruss2D 69    11  32  1  1
115. element corotTruss2D 70    12  33  1  1
116. element corotTruss2D 71    13  34  1  1
117. element corotTruss2D 72    14  35  1  1
118. element corotTruss2D 73    15  36  1  1
119. element corotTruss2D 74    16  37  1  1
120. element corotTruss2D 75    17  38  1  1
121. element corotTruss2D 76    18  39  1  1
122. element corotTruss2D 77    19  40  1  1
123. recorder  Node -file displacement.out -time -nodeRange 1 40
     -dof 1 2 disp
124. recorder  Node -file force.out -time -nodeRange 1 40 -dof 1 2
     reaction
125. pattern Plain 100 Linear {
126.      sp    40    2   -6.0
127. }
128. system BandGeneral
129. numberer RCM
130. constraints Transformation
131. integrator LoadControl 0.001
132. test NormDispIncr 1.0e-8 200 2
```

```
133. algorithm Newton
134. analysis Static
135. analyze 1000
```

Running the model and plotting the result (see Fig. 17.14) is done by using the Matlab codes as follows:

```
1.  clear all
2.  close all
3.  clc
4.  load node.tcl
5.  load element.tcl
6.  load displacement.out
7.  step=1000;
8.  for i=1:length(node(:,1))
9.  cCoor(i,1)=node(i,2)+displacement(step,2*i);
10. cCoor(i,2)=node(i,3)+displacement(step,2*i+1);
11. end
12. figure (1)
13. axis equal
14. grid on
15. xlim([-1,8])
16. ylim([-6,1])
17. set(gca, 'Fontsize',18)
18. xlabel('X Coordinate [m]')
19. ylabel('Y Coordinate [m]')
20. hold on
21. for i=1:length(element(:,1))
22. plot([cCoor(element(i,2),1),cCoor(element(i,3),1)],[cCoor
    (element(i,2),2),cCoor(element(i,3),2)],'lineWidth',3)
23. end
```

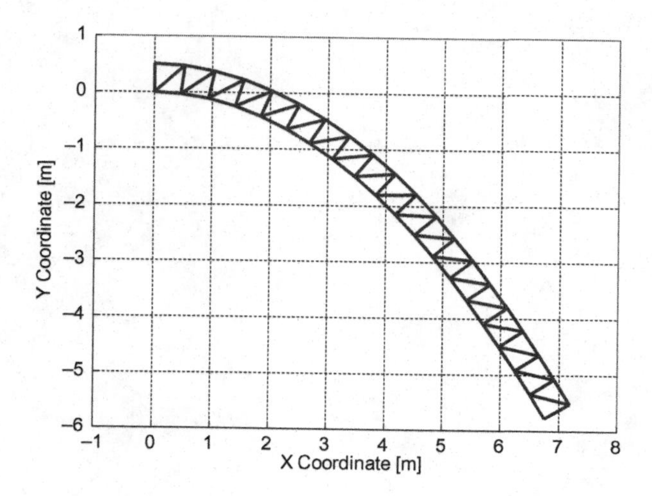

Fig. 17.14 Result of the model.

Reference

[1] Yaw, L. L. *2D Co-rotational Truss Formulation*. Walla Walla University, Washington, 2009.

Chapter 18

Implementing State-Based Peridynamics (SPD) in OpenSees

18.1 The Basic Theory of SPD

Similar to BPD, SPD discretizes the continuum body using a set of PD points (or PD nodes) [1]. Figures 18.1(a) and 18.1(b) represent the reference and current configurations of one PD point and the neighbor points in its horizon, e.g., $\mathbf{X}$ is the PD point in the reference configuration and $\mathbf{X}'$ is one of its neighboring points. $\mathbf{x}$ and $\mathbf{x}'$ are the current positions of $\mathbf{X}$ and $\mathbf{X}'$, respectively. $\boldsymbol{\xi} = \mathbf{X}' - \mathbf{X}$ is the bond vector between $\mathbf{X}$ and $\mathbf{X}'$. $\mathrm{H}_{\mathbf{X}}$ is the horizon (i.e., influence area) of the PD point $\mathbf{X}$. In the two-dimensional (2D) case, a horizon can be represented by a circular area centered at $\mathbf{X}$ with radius δ as shown in Fig. 18.1(a). The equation of motion for SPD point $\mathbf{X}$ has the following format (similar to Eq. (12.1)):

$$\rho \ddot{\mathbf{u}}(\mathbf{X}) = \int_{\mathrm{H}_{\mathbf{X}}} \left(\underline{\mathbf{T}}(\mathbf{X}) \langle \boldsymbol{\xi} \rangle - \underline{\mathbf{T}}(\mathbf{X}') \langle -\boldsymbol{\xi} \rangle \right) dV_{\mathbf{X}'} + \mathbf{b}(\mathbf{X}) \tag{18.1}$$

where $\ddot{\mathbf{u}}(\mathbf{X})$ is the acceleration vector of point $\mathbf{X}$, $\mathbf{b}(\mathbf{X})$ is the body force density of $\mathbf{X}$, $\underline{\mathbf{T}}$ is the force state field, which is the force density per unit volume squared, $\underline{\mathbf{T}}(\mathbf{X}) \langle \boldsymbol{\xi} \rangle$ is the force state of point $\mathbf{X}$ operating on $\boldsymbol{\xi}$, similarly, $\underline{\mathbf{T}}(\mathbf{X}') \langle -\boldsymbol{\xi} \rangle$ is the force state of point $\mathbf{X}'$ operating on $-\boldsymbol{\xi}$ (i.e., $\mathbf{X} - \mathbf{X}'$) and $V_{\mathbf{X}}$ and $V_{\mathbf{X}'}$ are the volumes of points $\mathbf{X}$ and $\mathbf{X}'$, respectively.

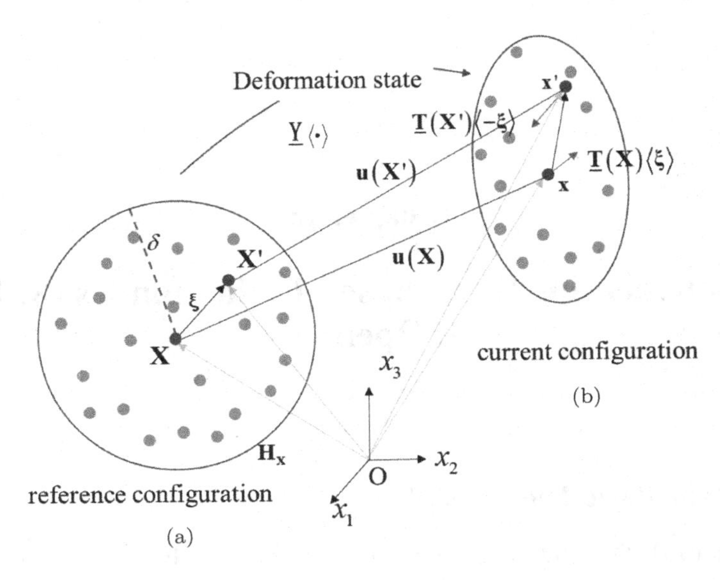

Fig. 18.1 Deformation of points in PD.

In order to calculate $\underline{\mathbf{T}}(\mathbf{X})\langle\boldsymbol{\xi}\rangle$, the shape tensors in the reference configuration $\mathbf{K}(\mathbf{X})$ and the current configuration $\mathbf{N}(\mathbf{X})$ must first be calculated as follows:

$$\mathbf{K}(\mathbf{X}) = \int_{\mathbf{H_X}} \underline{w}\langle\boldsymbol{\xi}\rangle(\boldsymbol{\xi}\otimes\boldsymbol{\xi})\,dV_{\boldsymbol{\xi}} \tag{18.2}$$

$$\mathbf{N}(\mathbf{X}) = \int_{\mathbf{H_X}} \underline{w}\langle\boldsymbol{\xi}\rangle\left(\boldsymbol{\xi} + \left(\mathbf{u}(\mathbf{X}') - \mathbf{u}(\mathbf{X})\right)\otimes\boldsymbol{\xi}\right)dV_{\boldsymbol{\xi}} \tag{18.3}$$

The symbol $\otimes$ represents the dyadic product of two vectors. $\mathbf{u}(\mathbf{X}')$ and $\mathbf{u}(\mathbf{X})$ are the displacement vectors of points $\mathbf{X}'$ and $\mathbf{X}$, respectively. $\underline{w}$ is a scalar state called the weight function describing relative importance between neighboring points. $\underline{w}$ can take the following form:

$$\underline{w}\langle\boldsymbol{\xi}\rangle = \begin{cases} \frac{2}{3} - 4*(\|\boldsymbol{\xi}\|/\delta)^2 + 4*(\|\boldsymbol{\xi}\|/\delta)^3 & \text{for } 0 \le \|\boldsymbol{\xi}\|/\delta < 1/2 \\ 4/3*(1-\|\boldsymbol{\xi}\|/\delta)^3 & \text{for } 1/2 \le \|\boldsymbol{\xi}\|/\delta \le 1 \end{cases} \tag{18.4}$$

The deformation gradient at point $\mathbf{X}$, $\mathbf{F}(\mathbf{X})$, can be obtained from

$$\mathbf{F}(\mathbf{X}) = \mathbf{N}(\mathbf{X})\,\mathbf{K}^{-1}(\mathbf{X}) \tag{18.5}$$

According to the small strain theory in continuum mechanics, the strain at point **X** is

$$\varepsilon(\mathbf{X}) = \frac{1}{2}\left(\mathbf{F}^T(\mathbf{X}) + \mathbf{F}(\mathbf{X})\right) - \mathbf{I} \tag{18.6}$$

where **I** is the identity matrix. The Cauchy stress can be calculated using the constitutive law as

$$\boldsymbol{\sigma}(\mathbf{X}) = \boldsymbol{\sigma}(\varepsilon(\mathbf{X})) \tag{18.7}$$

The first Piola–Kirchhoff stress of point **X** is then obtained [1]:

$$\mathbf{P}(\mathbf{X}) = \det(\mathbf{F}(\mathbf{X}))\,\boldsymbol{\sigma}(\mathbf{X})\,\mathbf{F}^{-T}(\mathbf{X}) \tag{18.8}$$

where $\det(\mathbf{F}(\mathbf{X}))$ is the determinate of $\mathbf{F}(\mathbf{X})$. The force vector state $\underline{\mathbf{T}}(\mathbf{X})\langle\boldsymbol{\xi}\rangle$ can be expressed as

$$\underline{\mathbf{T}}(\mathbf{X})\langle\boldsymbol{\xi}\rangle = \underline{w}\langle\boldsymbol{\xi}\rangle\,\mathbf{P}(\mathbf{X})\,\mathbf{K}^{-1}(\mathbf{X})\,\boldsymbol{\xi} \tag{18.9}$$

Similar to BPD, the equilibrium equation of SPD theory in Eq. (18.1) can be rewritten as a format of FEM as in Eq. (15.1), which can then be discretized and solved using an existing method in OpenSees, e.g., discretized by an implicit time-stepping method and solved by Newton Raphson solution algorithm. The details can be found in Section 15.3.

18.2 Implementation of SPD in the Framework of OpenSees

In OpenSees, as shown in Fig. 18.2, the **Domain** class stores the information of the FEM model, e.g., nodes (**Node** class), elements (**Element** class) and loadings (**Load** class). **Node** class stores nodal coordinates and responses (such as displacement, velocity and acceleration) and provides interfaces for other classes to operate these data. Considering the difference between PD material point and FEM node, it is necessary to add a new subclass, called **PDNode**, inheriting the **Node** class.

The **PDNode** class inherits all the public functions and variables of the **Node** class. It stores the information of the PD material point, such as the displacements, internal forces and coordinates, which are consistent with the functionality of the FE Node. Apart from this, the **PDNode** class also stores variables and functions of PD, such as its volume, neighbor points within horizon in both initial and current configurations,

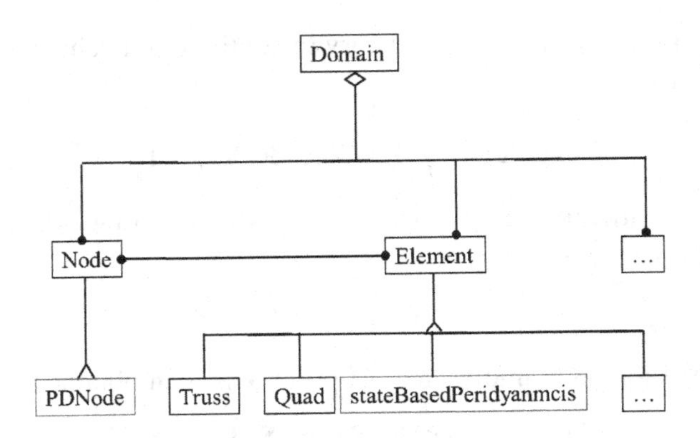

Fig. 18.2 Adding PDNode class and stateBasedPeridynamics class in OpenSees.

the deformation gradient (see Eq. (18.5)), the strain (see Eq. (18.6)), the stress (see Eqs. (18.7) and (18.8)), force states (see Eq. (18.9)), the influence function (see Eq. (18.4)) and the shape tensors under initial and current configurations (see Eqs. (18.2) and (18.3)). Furthermore, the **PDNode** class provides interfaces for other classes to get relevant information of the specified PD points for calculation.

The **Element** class in OpenSees performs element-level operations. There are several types of elements in OpenSees, such as truss element and four-node element. In this chapter, the SPD is implemented as a special type of element, i.e., adding a subclass (**stateBasedPeridynamics** class) of the **Element** class (like a macro element) in the framework of OpenSees. The role of the SPD element is to (1) calculate the strain of the PD point according to the displacement field and "send" the strain to the material level, (2) get the stress from the material level and calculate the internal force, and (3) calculate the stiffness matrix of the element. To implement the SPD theory in OpenSees, we need to implement the algorithms of calculating strain, internal force and stiffness in stateBasedPeridynamics class. Note that the stress can be calculated by using a 3D material model, which is independent of the element.

18.2.1 *PDNode class*

First, add blank files under the folder "node" in project "domain" and name them PDNode.cpp and PDNode.h (see Fig. 18.3). Enter the following code in PDNode.h:

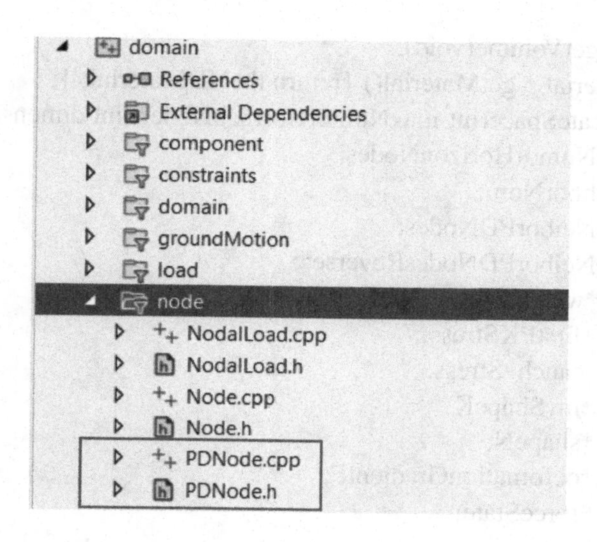

Fig. 18.3 Adding PDNode.cpp and PDNode.h in the folder "node".

```
1. #ifndef pdNode_h
2. #define pdNode_h
3. #include <DomainComponent.h>
4. #include <Node.h>
5. #include <NDMaterial.h>
6. class Element;
7. class Vector;
8. class Matrix;
9. class Channel;
10. class Renderer;
11. class DOF_Group;
12. class NodalThermalAction;
13. class PDNode : public Node
14. {
15. public:
16. PDNode(int tag, int ndof, double Crd1, double Crd2, NDMaterial
    &theNDMat, double vol, Vector *displayLoc=0 );
17. PDNode(int tag, int ndof, double Crd1, double Crd2, double Crd3,
    NDMaterial &theNDMat, double vol, Vector *displayLoc=0);
18. ~PDNode();
```

```
19. double getVolume(void);
20. NDMaterial * getMaterial() {return theNDMaterial; };
21. int allocateSpace(int maxNumofHorizonNodes, int dimension);
22. int maxNumofHorizonNodes;
23. int neighborNum;
24. ID *theNeiborPDNodes;
25. ID *theNeiborPDNodesReverse;
26. Vector *weightFun;
27. Matrix *firstPKStress;;
28. Vector *cauchyStress;
29. Matrix *invShapeK;
30. Matrix *shapeN;
31. Matrix *defomationGradientF;
32. Matrix *forceState;
33. Vector **xi;
34. Matrix *shapeK;
35. protected:
36. private:
37. double volume;
38. NDMaterial *theNDMaterial;
39. int pdNodeTag;
40. int dimension;
41. };
42. #endif
```

Lines 16–17 are constructors for the PD node in 2D and 3D models, respectively.

Line 20 is the material of the PD node.

In line 22, the variable maxNumofHorizonNodes represents the max number of neighbor particles within horizon.

In line 23, the variable "neighborNum" is the number of neighbor particles within horizon.

In line 24, theNeiborPDNodes is an array, which stores the tag of the neighbor point within horizon. For example, as shown in Fig. 18.4, #5 is the PD point, and #6, #7, #9, #8 and #12 are the neighbor nodes of #5. In this case, theNeiborPDNodes are [6, 7, 9, 8, 12].

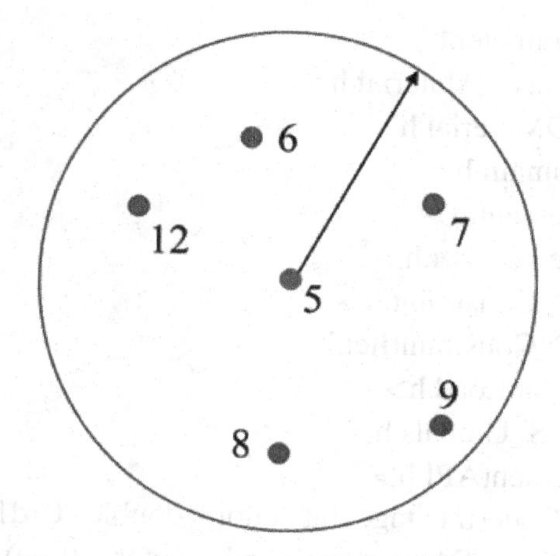

Fig. 18.4 The neighbor points of PD point #5.

In line 25, theNeiborPDNodesReverse is an array, which will be introduced later.

In line 26, weightFun is a vector which stores the value of weight function for the PDnode.

In line 27, firstPKStress is a vector which stores the first PK stress, and in line 28, cauchyStress is also a vector that stores the Cauchy stress.

The variables in lines 29–34 will be introduced in the stateBasedPeridynamics class later.

PDNode.h defines the variables and functions of **PDNode** class. The functions are implemented in pdnode.cpp, i.e.,

```
 1. #include <PDNode.h>
 2. #include <stdlib.h>
 3. #include <Element.h>
 4. #include <Vector.h>
 5. #include <Matrix.h>
 6. #include <Channel.h>
 7. #include <FEM_ObjectBroker.h>
 8. #include <DOF_Group.h>
 9. #include <Renderer.h>
10. #include <string.h>
11. #include <Information.h>
```

```
12. #include <Parameter.h>
13. #include <UniaxialMaterial.h>
14. #include <NDMaterial.h>
15. #include <Domain.h>
16. #include <Element.h>
17. #include <ElementIter.h>
18. #include <SP_Constraint.h>
19. #include <SP_ConstraintIter.h>
20. #include <NodalLoad.h>
21. #include <OPS_Globals.h>
22. #include <elementAPI.h>
23. PDNode::PDNode(int tag, int ndof, double Crd1, double Crd2,
    NDMaterial &theNDMat, double vol, Vector *dLoc)
24. :Node(tag, ndof, Crd1, Crd2, dLoc), volume(vol), theNDMaterial(0),
    pdNodeTag(tag)
25. {
26. theNDMaterial = theNDMat.getCopy("PlaneStrain");
27. if (theNDMaterial == 0) {
28. opserr ≪"FATAL PDNode::PDNode - failed to get a copy of material
    with tag " ≪tag ≪ endln;
29. exit(-1);
30. }
31. dimension = 2;
32. maxNumofHorizonNodes = 64;
33. neighborNum = -1;
34. this-> allocateSpace(maxNumofHorizonNodes, dimension);
35. }
36. PDNode::PDNode(int tag, int ndof, double Crd1, double Crd2, double
    Crd3, NDMaterial &theNDMat, double vol, Vector *dLoc)
37. :Node(tag, ndof, Crd1, Crd2, Crd3, dLoc), volume(vol), theNDMate-
    rial(0), pdNodeTag(tag)
38. {
39. theNDMaterial = theNDMat.getCopy("ThreeDimensional");
40. if (theNDMaterial == 0) {
41. opserr ≪"FATAL PDNode::PDNode - failed to get a copy of material
    with tag " ≪tag ≪ endln;
```

```
42. exit(-1);
43. }
44. dimension = 3;
45. maxNumofHorizonNodes = 216;
46. neighborNum = -1;
47. this-> allocateSpace(maxNumofHorizonNodes, dimension);
48. }
49. PDNode::~PDNode()
50. {
51. if (theNeiborPDNodes != 0)
52. delete theNeiborPDNodes;
53. if (theNeiborPDNodesReverse != 0)
54. delete theNeiborPDNodesReverse;
55. if (shapeK != 0)
56. delete shapeK;
57. if (invShapeK != 0)
58. delete invShapeK;
59. if (shapeN != 0)
60. delete shapeN;
61. if (defomationGradientF != 0)
62. delete defomationGradientF;
63. if (weightFun != 0)
64. delete weightFun;
65. }
66. int
67. PDNode :: allocateSpace(int maxNumofHorizonNodes, int dimen-
    sion) {
68. weightFun = new Vector(maxNumofHorizonNodes);
69. weightFun-> Zero();
70. theNeiborPDNodes = new ID(maxNumofHorizonNodes);
71. theNeiborPDNodes-> Zero();
72. theNeiborPDNodesReverse = new ID(maxNumofHorizonNodes);
73. theNeiborPDNodesReverse-> Zero();
74. shapeK = new Matrix(dimension, dimension);
75. shapeK-> Zero();
76. invShapeK = new Matrix(dimension, dimension);
```

```
77. invShapeK-> Zero();
78. shapeN = new Matrix(dimension, dimension);
79. shapeN-> Zero();
80. defomationGradientF = new Matrix(dimension, dimension);
81. defomationGradientF-> Zero();
82. firstPKStress = new Matrix(dimension, dimension);
83. firstPKStress->Zero();
84. cauchyStress = new Vector(dimension * 3 - 3);
85. cauchyStress-> Zero();
86. forceState = new Matrix(maxNumofHorizonNodes, dimension);
87. forceState-> Zero();
88. xi = new Vector * [maxNumofHorizonNodes];
89. return 0;
90. }
91. double
92. PDNode::getVolume()
93. {
94. return volume;
95. }
```

Lines 1–22 include the necessary head files in the class; Lines 23–35 and lines 36–65 are constructors which are used to create two-dimensional and three-dimensional PD nodes, respectively. Lines 66–90 are allocateSpace functions used to initialize variables (set the dimensions of a matrix or vector and initialize the value to **0**). Lines 92–96 create a function to get the volume of the PD node.

The tcl command used to create a PD point in Tcl is

> *pdnode $nodeTag $coords $matTag $volume <-mass $massValue> <-dispLoc $displacementLocationValue> <-disp $dispValue> <-vel $velocityValue>*

where *$nodeTag* is the tag of the PD node, *$coords* is the nodal coordinates vector of the PD node, if the model is 2D, *$coords* consists of coordinates of x and y, if the model is 3D, *$coords* consists of coordinates of x, y and z, *$matTag* presents the material tag, which is used to calculate the stress of the PD node, *$vol* is the bulk volume of the PD node; *$massValue* is the nodal mass vector, *$displacementLocationValue* is the nodal displacement location vector, *$dispValue* is the nodal displacement vector of the PD node and *$velocityValue* is the nodal velocity vector of the PD node.

To interpret the *pdnode* in OpenSees, it is necessary to add a function called "TclCommand_addPDNode" in TclModelBuilder.cpp:

```
1. int
2. TclCommand_addPDNode(ClientData clientData, Tcl_Interp*
   interp, int argc,
3. TCL_Char** argv)
4. {
5. if (theTclBuilder == 0) {
6. opserr ≪"WARNING builder has been destroyed" ≪endln;
7. return TCL_ERROR;
8. }
9. int ndm = theTclBuilder->getNDM();
10. int ndf = theTclBuilder->getNDF();
11. if (argc <3 + ndm) {
12. opserr ≪"WARNING insufficient arguments\n";
13. printCommand(argc, argv);
14. opserr ≪"Want: node nodeTag? [ndm coordinates?] <-mass [ndf
    values?]>\n";
15. return TCL_ERROR;
16. }
17. PDNode* thePDNode = 0;
18. int nodeId;
19. if (Tcl_GetInt(interp, argv[1], &nodeId) != TCL_OK) {
20. opserr ≪"WARNING invalid nodeTag\n";
21. opserr ≪"Want: node nodeTag? [ndm coordinates?] <-mass [ndf
    values?]> \n";
22. return TCL_ERROR;
23. }
24. double xLoc, yLoc, zLoc;
25. int matTag;
26. double vol;
27. UniaxialMaterial* theUniaxialMat;
28. NDMaterial* theNDMat;
```

```
29. if (ndm == 2) {
30. if (Tcl_GetDouble(interp, argv[2], &xLoc) != TCL_OK) {
31. opserr <<"WARNING invalid XCoordinate\n";
32. opserr <<"node: " <<nodeId <<endln;
33. return TCL_ERROR;
34. }
35. if (Tcl_GetDouble(interp, argv[3], &yLoc) != TCL_OK) {
36. opserr <<"WARNING invalid YCoordinate\n";
37. opserr <<"node: " <<nodeId <<endln;
38. return TCL_ERROR;
39. }
40. if (Tcl_GetInt(interp, argv[4], &matTag) != 0) {
41. opserr <<"WARNING: Invalid matTag for PDNode" <<nodeId
    <<"\n";
42. return TCL_ERROR;
43. }
44. if (Tcl_GetDouble(interp, argv[5], &vol) != TCL_OK) {
45. opserr <<"WARNING invalid vol\n";
46. opserr <<"node: " <<nodeId <<endln;
47. return TCL_ERROR;
48. }
49. theNDMat = OPS_getNDMaterial(matTag);
50. }
51. else if (ndm == 3) {
52. if (Tcl_GetDouble(interp, argv[2], &xLoc) != TCL_OK) {
53. opserr <<"WARNING invalid XCoordinate\n";
54. opserr <<"node: " <<nodeId <<endln;
55. return TCL_ERROR;
56. }
57. if (Tcl_GetDouble(interp, argv[3], &yLoc) != TCL_OK) {
58. opserr <<"WARNING invalid YCoordinate\n";
59. opserr <<"node: " <<nodeId <<endln;
60. return TCL_ERROR;
61. }
62. if (Tcl_GetDouble(interp, argv[4], &zLoc) != TCL_OK) {
63. opserr <<"WARNING invalid ZCoordinate\n";
```

```
64. opserr ≪"node: " ≪nodeId ≪endln;
65. return TCL_ERROR;
66. }
67. if (Tcl_GetInt(interp, argv[5], &matTag) != 0) {
68. opserr ≪"WARNING: Invalid matTag for PDNode" ≪nodeId
    ≪"\n";
69. return TCL_ERROR;
70. }
71. if (Tcl_GetDouble(interp, argv[6], &vol) != TCL_OK) {
72. opserr ≪"WARNING invalid vol\n";
73. opserr ≪"node: " ≪nodeId ≪endln;
74. return TCL_ERROR;
75. }
76. theNDMat = OPS_getNDMaterial(matTag);
77. }
78. else {
79. opserr ≪"WARNING invalid ndm\n";
80. opserr ≪"node: " ≪nodeId ≪endln;;
81. return TCL_ERROR;
82. }
83. int currentArg = 4 + ndm;
84. if (currentArg <argc && strcmp(argv[currentArg], "-ndf") == 0) {
85. if (Tcl_GetInt(interp, argv[currentArg + 1], &ndf) != TCL_OK) {
86. opserr ≪"WARNING invalid nodal ndf given for node " ≪nodeId
    ≪endln;
87. return TCL_ERROR;
88. }
89. currentArg += 2;
90. }
91. if (ndm == 2)
92. thePDNode = new PDNode(nodeId, ndf, xLoc, yLoc, *theNDMat,
    vol);
93. else
94. thePDNode = new PDNode(nodeId, ndf, xLoc, yLoc, zLoc,
    *theNDMat, vol);
```

```
95. if (theTclDomain->addNode(thePDNode) == false) { //notice it's
    addNode, not addPDNode
96. opserr ≪"WARNING failed to add node to the domain\n";
97. opserr ≪"node: " ≪nodeId ≪endln;
98. delete thePDNode; // otherwise memory leak
99. return TCL_ERROR;
100. }
101. while (currentArg <argc) {
102. if (strcmp(argv[currentArg], "-mass") == 0) {
103. currentArg++;
104. if (argc <currentArg + ndf) {
105. opserr ≪"WARNING incorrect number of nodal mass terms\n";
106. opserr ≪"node: " ≪nodeId ≪endln;
107. return TCL_ERROR;
108. }
109. Matrix mass(ndf, ndf);
110. double theMass;
111. for (int i = 0; i <ndf; i++) {
112. if (Tcl_GetDouble(interp, argv[currentArg++], &theMass) !=
     TCL_OK) {
113. opserr ≪"WARNING invalid nodal mass term\n";
114. opserr ≪"node: " ≪nodeId ≪", dof: " ≪i + 1 ≪endln;
115. return TCL_ERROR;
116. }
117. mass(i, i) = theMass;
118. }
119. thePDNode-> setMass(mass);
120. }
121. else if (strcmp(argv[currentArg], "-dispLoc") == 0) {
122. currentArg++;
123. if (argc <currentArg + ndm) {
124. opserr ≪"WARNING incorrect number of nodal display location
     terms, need ndm\n";
125. opserr ≪"node: " ≪nodeId ≪endln;
126. return TCL_ERROR;
127. }
```

```
128. Vector displayLoc(ndm);
129. double theCrd;
130. for (int i = 0; i <ndm; i++) {
131. if (Tcl_GetDouble(interp, argv[currentArg++], &theCrd) !=
     TCL_OK) {
132. opserr ≪"WARNING invalid nodal mass term\n";
133. opserr ≪"node: " ≪nodeId ≪", dof: " ≪i + 1 ≪endln;
134. return TCL_ERROR;
135. }
136. displayLoc(i) = theCrd;
137. }
138. thePDNode-> setDisplayCrds(displayLoc);
139. }
140. else if (strcmp(argv[currentArg], "-disp") == 0) {
141. currentArg++;
142. if (argc <currentArg + ndf) {
143. opserr ≪"WARNING incorrect number of nodal disp terms\n";
144. opserr ≪"node: " ≪nodeId ≪endln;
145. return TCL_ERROR;
146. }
147. Vector disp(ndf);
148. double theDisp;
149. for (int i = 0; i <ndf; i++) {
150. if (Tcl_GetDouble(interp, argv[currentArg++], &theDisp) !=
     TCL_OK) {
151. opserr ≪"WARNING invalid nodal disp term\n";
152. opserr ≪"node: " ≪nodeId ≪", dof: " ≪i + 1 ≪endln;
153. return TCL_ERROR;
154. }
155. disp(i) = theDisp;
156. }
157. thePDNode->setTrialDisp(disp);
158. thePDNode->commitState();
159. }
160. else if (strcmp(argv[currentArg], "-vel") == 0) {
161. currentArg++;
```

```
162. if (argc <currentArg + ndf) {
163. opserr ≪"WARNING incorrect number of nodal vel terms\n";
164. opserr ≪"node: " ≪nodeId ≪endln;
165. return TCL_ERROR;
166. }
167. Vector disp(ndf);
168. double theDisp;
169. for (int i = 0; i <ndf; i++) {
170. if (Tcl_GetDouble(interp, argv[currentArg++], &theDisp) !=
     TCL_OK) {
171. opserr ≪"WARNING invalid nodal vel term\n";
172. opserr ≪"node: " ≪nodeId ≪", dof: " ≪i + 1 ≪endln;
173. return TCL_ERROR;
174. }
175. disp(i) = theDisp;
176. }
177. thePDNode-> setTrialVel(disp);
178. thePDNode-> commitState();
179. }
180. else
181. currentArg++;
182. }
183. return TCL_OK;
184. }
```

18.2.2 *StateBasedPeridynamics class*

This section introduces the implementation of the **stateBasedPeridynamics** class. First, create a folder called "peridynamics" under the element project as shown in Fig. 18.5, add two blank files (you can also copy any element files and then modify them based on the files) and name them stateBasedPeridynamics.h and stateBasedPeridynamics.cpp.

Second, define the format of creating the element in the Tcl model as follows:

spd $elementTag $ndm $firstPDnode $lastPDnode $horizonSize $dx <-thickness $thicknessValue>

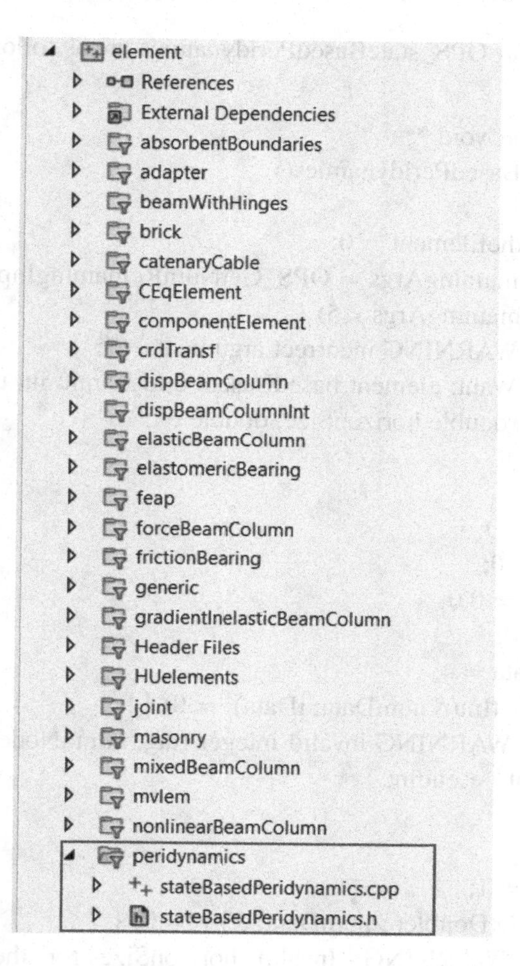

Fig. 18.5 Add the **stateBasedPeridynamics** class in project element.

where *$elementTag* is the tag of the element, *$ndm* is the dimension, *$firstPDnode* is the first number of the PD node, *$lastPDnode* is the last number of the PD node, *$horizonSize* is the size of horizon and *$dx* is the grid size of the model. If it is a 2D problem, thickness is required and *$thicknessValue* is the thickness of the 2D model. It is noted that the tags of the PD node must be consecutive for simplicity.

Third, a function called "OPS_stateBasedPeridynamics" is added in TclElementCommand.cpp to interpret the TCL command "*spd*".

The codes for OPS_stateBasedPeridynamics are as follows:

```
1. OPS_Export void *
2. OPS_stateBasedPeridynamics()
3. {
4. Element *theElement = 0;
5. int numRemainingArgs = OPS_GetNumRemainingInputArgs();
6. if (numRemainingArgs <5) {
7. opserr <<"WARNING incorrect arguments\n";
8. opserr <<"Want: element basedBasedPeridynamic int tag, int ndFirst,
   int ndLast, double horizonSize, double dx; \n";
9. return 0;
10. }
11. double h = 0;
12. double r = 0;
13. double dx = 0.0;
14. int iData[4];
15. int numData = 4;
16. if (OPS_GetInt(&numData, iData) != 0) {
17. opserr <<"WARNING invalid integer (tag, ndm,iNode, jNode) in the
    pd element" <<endln;
18. return 0;
19. }
20. numData = 1;
21. if (OPS_GetDouble(&numData, &h) != 0) {
22. opserr <<"WARNING: Invalid horizonSize for the pd element"
    <<iData[0] <<" $iNode $jNode $horizonSize $r\n";
23. return 0;
24. }
25. numData = 1;
26. if (OPS_GetDouble(&numData, &dx) != 0) {
27. opserr <<"WARNING: Invalid horizonSize for the pd element"
    <<iData[0] <<" $iNode $jNode $horizonSize $r\n";
28. return 0;
29. }
30. double thickness = dx;
```

```
31. while (OPS_GetNumRemainingInputArgs() > 0) {
32. const char* type = OPS_GetString();
33. if (strcmp(type, "-thickness") == 0) {
34. if (OPS_GetNumRemainingInputArgs() <1) {
35. opserr <<"WARNING invalid parameter -thickness $thickness\n";
36. return 0;
37. }
38. else {
39. if (OPS_GetDoubleInput(&numData, &thickness) <0) {
40. opserr <<"WARNING invalid parameter :: no thickness info\n";
41. return 0;
42. }
43. }
44. }
45. }
46. theElement = new stateBasedPeridynamics(iData[0], iData[1],
    iData[2], iData[3], h, dx, thickness);
47. if (theElement == 0) {
48. opserr <<"WARNING: out of memory: state pd element " <<iData[0]
    <<"\n";
49. }
50. return theElement;
51. }
```

To implement the PD theory, the variables and functions are defined in the stateBasedPeridynamics.h at first, and then the functions are implemented in stateBasedPeridynamics.cpp. Some major functions are introduced as follows:

The function "getNodePtrs" is used to get the pointers to thePDNodes array:

```
1. Node **
2. stateBasedPeridynamics::getNodePtrs(void)
3. {
4. Node ** theNodes = (Node**)thePDNodes;
5. return theNodes;
6. }
```

The function "setDomain" is used to add the element to the domain. For the sake of convenience, the relationship of each PD node and its neighboring points within horizon is also established in this function, as well as the weight functions (refer to Eq. (18.4)) and the shape function **K** in the initial configuration (refer to Eq. (18.2)):

```
1. void
2. stateBasedPeridynamics::setDomain(Domain *theDomain)
3. {
4. if (thePDNodes ==0) thePDNodes = new PDNode *[numExternalN-
   odes];
5. for (int i = 0; i <numExternalNodes; i++) {
6. thePDNodes[i] = theDomain -> getPDNode(connected ExternalN-
   odes(i)); //notice it's getPDNode...
7. if (thePDNodes[i] == 0) {
8. opserr ≪"WARNING:stateBasedPeridynamics( ) - node: ";
9. opserr ≪connectedExternalNodes(i) ≪" does not exist ! " ≪endln;
10. exit(-1);
11. }
12. }
13. this->DomainComponent::setDomain(theDomain);
14. this->updatedHorizon();
15. this->calculateWeight();
16. this->calculateShapeFunK();
17. }
```

The function "update" is one of the most important functions in the element. It calculates the strain of the element. In the finite element method, the strain at the Gauss point is calculated by using the displacement–strain transformation matrix, while in SPD, the strain at the PD point is calculated using the displacement of PD points within its horizon, i.e., using Eqs. (18.2)–(18.6):

```
1. int
2. stateBasedPeridynamics::update()
3. {
4. int err = 0;
```

```
5. static Matrix tempFTranspose(dimension, dimension);
6. static Vector vectorE(dimension * 3 - 3);
7. static Matrix strainE(dimension, dimension);
8. eta->Zero();
9. dispIJ.Zero();
10. for (int i = 0; i <numExternalNodes; i++) {
11. thePDNodes[i]->shapeN->Zero();
12. temp->Zero();
13. thePDNodes[i]->defomationGradientF->Zero();
14. strainE.Zero();
15. vectorE.Zero();
16. tempTrialDispI.Zero();
17. tempTrialDispJ.Zero();
18. for (int j = 0; j <= thePDNodes[i]->neighborNum; j++) {
19. const Vector &trialDispI = thePDNodes[i]->getTrialDisp();
20. const Vector &trialDispJ = thePDNodes[(*(thePDNodes[i]->the
    NeiborPDNodes))(j)]->getTrialDisp();
21. tempTrialDispI(0) = trialDispI(0) + (*perturb[i])(0);
22. tempTrialDispI(1) = trialDispI(1) + (*perturb[i])(1);
23. tempTrialDispJ(0) = trialDispJ(0) + (*perturb[(*(thePDNodes[i]-
    >theNeiborPDNodes))(j)])(0);
24. tempTrialDispJ(1) = trialDispJ(1) + (*perturb[(*(thePDNodes[i]-
    >theNeiborPDNodes))(j)])(1);
25. if (dimension == 3) {
26. tempTrialDispI(2) = trialDispI(2) + (*perturb[i])(2);
27. tempTrialDispJ(2) = trialDispJ(2) + (*perturb[(*(thePDNodes[i]-
    >theNeiborPDNodes))(j)])(2);
28. }
29. dispIJ = tempTrialDispJ - tempTrialDispI;
30. *eta = *(thePDNodes[i]->xi)[j] + dispIJ;
31. (*temp)(0, 0) = dispIJ(0)*(*(thePDNodes[i]->xi)[j])(0);
32. (*temp)(0, 1) = dispIJ(0)*(*(thePDNodes[i]->xi)[j])(1);
33. (*temp)(1, 0) = dispIJ(1)*(*(thePDNodes[i]->xi)[j])(0);
34. (*temp)(1, 1) = dispIJ(1)*(*(thePDNodes[i]->xi)[j])(1);
35. f (dimension == 3) {
36. (*temp)(0, 2) = dispIJ(0)*(*(thePDNodes[i]->xi)[j])(2);
```

```
37. (*temp)(1, 2) = dispIJ(1)*(*(thePDNodes[i]->xi)[j])(2);
38. (*temp)(2, 0) = dispIJ(2)*(*(thePDNodes[i]->xi)[j])(0);
39. (*temp)(2, 1) = dispIJ(2)*(*(thePDNodes[i]->xi)[j])(1);
40. (*temp)(2, 2) = dispIJ(2)*(*(thePDNodes[i]->xi)[j])(2);
41. }
42. thePDNodes[i]->shapeN->addMatrix(1.0, *temp, thePDNodes[(*(the
    PDNodes[i]->theNeiborPDNodes))(j)]->getVolume()*(*(thePDNod
    es[i]->weightFun))(j));
43. }
44. theNodes[i]->defomationGradientF->addMatrixProduct(0.0,
    *(thePDNodes[i]->shapeN), *(thePDNodes[i]->invShapeK), 1.0);
45. thePDNodes[i]->defomationGradientF->addMatrix(1.0, *ones, 1.0);
46. tempFTranspose.Zero();
47. tempFTranspose.addMatrixTranspose(0.0, *(thePDNodes[i]->defoma
    tionGradientF), 1.0);
48. tempFTranspose.addMatrix(0.5, *(thePDNodes[i]->defomationGradi
    entF), 0.5);
49. strainE.addMatrix(0.0, tempFTranspose, 1.0);
50. strainE.addMatrix(1.0, *ones, -1.0);
51. if (dimension == 2) {
52. vectorE(0) = strainE(0, 0);
53. vectorE(1) = strainE(1, 1);
54. vectorE(2) = strainE(0, 1) + strainE(1, 0);
55. }
56. else if (dimension == 3) {
57. vectorE(0) = strainE(0, 0);
58. vectorE(1) = strainE(1, 1);
59. vectorE(2) = strainE(2, 2);
60. vectorE(3) = strainE(0, 1) + strainE(1, 0);
61. vectorE(4) = strainE(0, 2) + strainE(2, 0);
62. vectorE(5) = strainE(1, 2) + strainE(2, 1);
63. }
64. thePDNodes[i]->getMaterial()->setTrialStrain(vectorE);
65. err = 0;
66. }
67. if (err!=0) {
68. opserr <<"error in setting strain at the PD node" <<endln;
```

```
69. }
70. return 0;
71. }
```

Lines 18–42 compute the shape function N in the current configuration (refer to Eq. (18.3)). Lines 43–44 calculate the deformation gradient (refer to Eq. (18.5)). Lines 45–50 calculate the strain at the PD point by small strain theory (refer to Eq. (18.6)). Lines 51–64 convert the matrix of strain to vector and send it to the material to calculate the stress.

The "getTangentStiff" function of the element is used to calculate the element stiffness. Since it is difficult to derive and implement the consistent tangent of SPD, here we obtain the stiffness by using the perturbation method:

```
1. const Matrix & stateBasedPeridynamics::getTangentStiff(void)
2. {
3. if (theTangent == 0) {
4. theTangent = new Matrix(numExternalNodes*dimension,
   numExternalNodes*dimension);
5. }
6. theTangent->Zero();
7. staticVector
   tempResistingForce(numExternalNodes*dimension);
8. for (int i = 0; i <numExternalNodes*dimension; i++) {
9. tempResistingForce(i) = (*P)(i);
10. }
11. double perturbValue = 1e-11;
12. for (int i = 0; i <numExternalNodes; i++) {
13. perturb[i]->Zero();
14. const Vector &trialDispI = thePDNodes[i]->getTrialDisp();
15. for (int m = 0; m <dimension; m++) {
16. if (abs(trialDispI(m)) >1e-6) {
17. perturbValue = trialDispI(m)*1e-3;
18. }
19. (*perturb[i])(m) + = perturbValue;
20. this->update();
21. this->getResistingForce();
```

```
22. for (int j = 0; j <numExternalNodes*dimension; j++) {
23. (*theTangent)(j, dimension*i + m) = ((*P)(j) - tempResisting-
    Force(j))/ perturbValue;
24. }
25. perturb[i]->Zero();
26. }
27. this->update();
28. this->getResistingForce();
29. Ki = theTangent;
30. }
31. return *theTangent;
32. }
```

The function, "getResistingForce" of the element, is used to calculate the resistance force of the element (refer to Eqs. (18.1) and (18.9)). The codes are as follows:

```
 1. const Vector & stateBasedPeridynamics::getResistingForce(void)
 2. {
 3. this->calculateForceStateT();
 4. if (P == 0) {
 5. P = new Vector(numExternalNodes*dimension);
 6. }
 7. P->Zero();
 8. static Vector tempP(dimension);
 9. static Vector tempForceStateTij(dimension);
10. static Vector tempForceStateTji(dimension);
11. static Vector tempRealP(dimension);
12. for (int i = 0; i <numExternalNodes; i++) {
13. tempForceStateTij.Zero();
14. tempForceStateTji.Zero();
15. tempP.Zero();
16. tempRealP.Zero();
17. for (int j = 0; j <= thePDNodes[i]->neighborNum; j++) {
18. tempForceStateTij(0) = (*(thePDNodes[i]->forceState))(j, 0);
19. tempForceStateTij(1) = (*(thePDNodes[i]->forceState))(j, 1);
```

```
20. tempForceStateTji(0) = (*(thePDNodes[(*(thePDNodes[i]-
    >theNeiborPDNodes))(j)]->forceState))((*(thePDNodes[i]-
    >theNeiborPDNodesReverse))(j), 0);
21. tempForceStateTji(1) = (*(thePDNodes[(*(thePDNodes[i]-
    >theNeiborPDNodes))(j)]->forceState))((*(thePDNodes[i]-
    >theNeiborPDNodesReverse))(j), 1);
22. if (dimension == 3) {
23. tempForceStateTij(2) = (*(thePDNodes[i]->forceState))(j, 2);
24. tempForceStateTji(2) = (*(thePDNodes[(*(thePDNodes[i]-
    >theNeiborPDNodes))(j)]->forceState))((*(thePDNodes[i]-
    >theNeiborPDNodesReverse))(j), 2);
25. }
26. tempRealP.addVector(1.0, tempForceStateTji - tempForceStateTij,
    1.0);
27. }
28. (*P)(i*dimension) = tempRealP(0);
29. (*P)(i*dimension + 1) = tempRealP(1);
30. if (dimension == 3) {
31. (*P)(i*dimension + 2) = tempRealP(2);
32. }
33. }
34. return *P;
35. }
```

Line 3 calls the function to calculate the force state of each PD point (refer to Eq. (18.9)). Lines 12–33 calculate the resistance force of the PD element based on the force state.

The function "updatedHorizon" established the relationship between each PD point and its neighboring points within horizon:

```
1. void
2. stateBasedPeridynamics::updatedHorizon() {
3. for (int i = 0; i <numExternalNodes-1; i++){
4. for (int j = i+1; j <numExternalNodes; j++){
5. tempCrdJ.Zero();
6. const Vector &crdsI = thePDNodes[i]->getCrds();
7. const Vector &crdsJ = thePDNodes[j]->getCrds();
```

```
8. if ((crdsI - crdsJ).Norm() <horizonSize ) {
9. thePDNodes[i]->neighborNum = thePDNodes[i]->neighborNum
   + 1;
10. thePDNodes[j]->neighborNum = thePDNodes[j]->neighborNum
   + 1;
11. thePDNodes[i]->xi)[thePDNodes[i]->neighborNum] = new Vec-
   tor(dimension);
12. *(thePDNodes[i]->xi)[thePDNodes[i]->neighborNum] = crdsJ -
   crdsI;
13. (thePDNodes[j]->xi)[thePDNodes[j]->neighborNum] = new
   Vector(dimension);
14. *(thePDNodes[j]->xi)[thePDNodes[j]->neighborNum] = crdsI -
   crdsJ;
15. *(thePDNodes[i]->theNeiborPDNodes))(thePDNodes[i]-
   >neighborNum) = j;
16. (*(thePDNodes[j]->theNeiborPDNodes))(thePDNodes[j]-
   >neighborNum) = i;
17. (*(thePDNodes[i]->theNeiborPDNodesReverse))(thePDNodes[i]-
   >neighborNum) = thePDNodes[j]->neighborNum;
18. (*(thePDNodes[j]->theNeiborPDNodesReverse))(thePDNodes[j]-
   >neighborNum) = thePDNodes[i]->neighborNum;
19. }
20. }
21. }
22. }
```

In lines 15–18, thePDNodes[i]->theNeiborPDNodes and thePDNodes [i]->theNeibor PDNodesReverse are important variables in the **stateBasedPeridynamics** class.

(*(thePDNodes[i]->theNeiborPDNodes))(k)=n represents that n is kth neighbor particle of PD point #i;

(*(thePDNodes[i]->theNeiborPDNodesReverse))(k)=m represents that #i is the mth neighbor particle of PD point n. For example, as shown in Fig. 18.6, the neighbor particles of #5 are [6, 7, 9, 8, 12], and 9 is the third neighbor particle of PD point #5, i.e.,

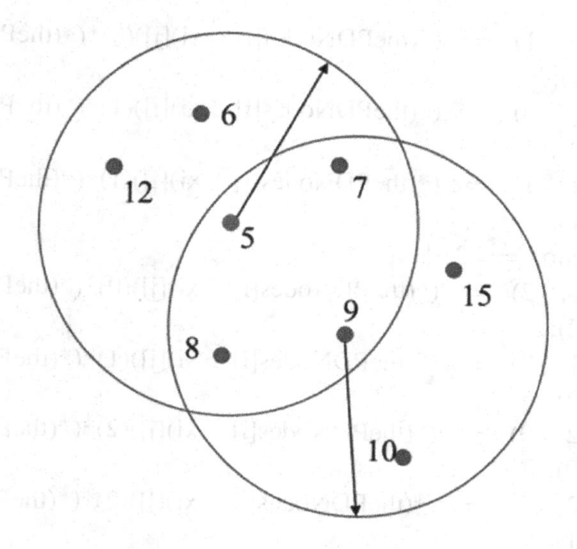

Fig. 18.6 The neighbor points of PD points #5 and #9.

(*(thePDNodes[4]->theNeiborPDNodes))(3)=9; the neighbor particles of #9 are [7, 5, 8, 10, 15], and #5 is the second neighbor particle of #9, i.e., (*(thePDNodes[4]->theNeiborPDNodesReverse))(3) = 2; it should be noted that the thePDNodes[4] represents the PD node #5 since the array starts from 0 in C++.

The function "calculateShapeFunK" calculates the shape function K and its inverse of each PD point, see Eq. (15.3):

```
1. void
2. stateBasedPeridynamics::calculateShapeFunK() {
3. eta->Zero();
4. for (int i = 0; i <numExternalNodes; i++) {
5. thePDNodes[i]->shapeK->Zero();
6. temp->Zero();
7. thePDNodes[i]->invShapeK->Zero();
8. for (int j = 0; j <= thePDNodes[i]->neighborNum; j++) {
9. (*temp)(0, 0) = (*(thePDNodes[i]->xi)[j])(0)*(*(thePDNodes[i]-
   >xi)[j])(0);
```

```
10. (*temp)(0, 1) = (*(thePDNodes[i]->xi)[j])(0)*(*(thePDNodes[i]-
    >xi)[j])(1);
11. (*temp)(1, 0) = (*(thePDNodes[i]->xi)[j])(1)*(*(thePDNodes[i]-
    >xi)[j])(0);
12. (*temp)(1, 1) = (*(thePDNodes[i]->xi)[j])(1)*(*(thePDNodes[i]-
    >xi)[j])(1);
13. if (dimension == 3) {
14. (*temp)(0, 2) = (*(thePDNodes[i]->xi)[j])(0)*(*(thePDNodes[i]-
    >xi)[j])(2);
15. (*temp)(1, 2) = (*(thePDNodes[i]->xi)[j])(1)*(*(thePDNodes[i]-
    >xi)[j])(2);
16. (*temp)(2, 0) = (*(thePDNodes[i]->xi)[j])(2)*(*(thePDNodes[i]-
    >xi)[j])(0);
17. (*temp)(2, 1) = (*(thePDNodes[i]->xi)[j])(2)*(*(thePDNodes[i]-
    >xi)[j])(1);
18. (*temp)(2, 2) = (*(thePDNodes[i]->xi)[j])(2)*(*(thePDNodes[i]-
    >xi)[j])(2);
19. }
20. thePDNodes[i]->shapeK->addMatrix(1.0, *temp, thePDNodes[(*(the
    PDNodes[i]->theNeiborPDNodes))(j)]->getVolume()*(*(thePD
    Nodes[i]->weightFun))(j));
21. }
22. thePDNodes[i]->shapeK->Invert(*(thePDNodes[i]->invShapeK));
23. }
24. }
```

The function “calculateForceStateT” is used to calculate the force state of each PD point, see Eq. (15.9):

```
1. void
2. stateBasedPeridynamics::calculateForceStateT() {
3. static Vector tempStress(3 * dimension - 3);
4. static Vector tempForceState(dimension);
5. static Vector tempForceState2(dimension);
6. static Matrix firstPKRealStress(dimension, dimension);
7. double pi = 3.14159265358979323846264338327950288419716939
   99;
```

```
 8. double detF;
 9. Matrix Ft(dimension, dimension);
10. Matrix invFt(dimension, dimension);
11. eta->Zero();
12. dispIJ.Zero();
13. for (int i = 0; i <numExternalNodes; i++) {
14. tempStress = thePDNodes[i]->getMaterial()->getStress();
15. if (dimension == 2) {
16. (*thePDNodes[i]->cauchyStress)(0) = tempStress(0);
17. (*thePDNodes[i]->cauchyStress)(1) = tempStress(1);
18. (*thePDNodes[i]->cauchyStress)(2) = tempStress(2);
19. }
20. else if (dimension == 3) {
21. (*thePDNodes[i]->cauchyStress)(0) = tempStress(0);
22. (*thePDNodes[i]->cauchyStress)(1) = tempStress(1);
23. (*thePDNodes[i]->cauchyStress)(2) = tempStress(2);
24. (*thePDNodes[i]->cauchyStress)(3) = tempStress(3);
25. (*thePDNodes[i]->cauchyStress)(4) = tempStress(4);
26. (*thePDNodes[i]->cauchyStress)(5) = tempStress(5);
27. }
28. }
29. for (int i = 0; i <numExternalNodes; i++) {
30. tempTrialDispI.Zero();
31. tempTrialDispJ.Zero();
32. firstPKRealStress.Zero();
33. thePDNodes[i]->forceState->resize
    (thePDNodes[i]->maxNumofHorizonNodes, dimension);
34. thePDNodes[i]->forceState->Zero();
35. tempForceState.Zero();
36. tempForceState2.Zero();
37. temp->Zero();
38. if (dimension == 2) {
39. (*temp)(0, 0) = (*thePDNodes[i]->cauchyStress)(0);
40. (*temp)(1, 1) = (*thePDNodes[i]->cauchyStress)(1);
41. (*temp)(0, 1) = (*thePDNodes[i]->cauchyStress)(2);
42. (*temp)(1, 0) = (*temp)(0, 1);
```

```
43. Ft.addMatrixTranspose(0.0,
    *(thePDNodes[i]->defomationGradientF), 1.0);
44. Ft.Invert(invFt);
45. detF = (*(thePDNodes[i]->defomationGradientF))(0, 0)*(*(thePD
    Nodes[i]->defomationGradientF))(1, 1) - (*(thePDNodes[i]->defom
    ationGradientF))(0, 1)*(*(thePDNodes[i]->defomationGradientF))
    (1, 0);
46. if (abs(detF) <1e-14) {
47. opserr <<" detF is zero, something wrong!" <<endln;
48. exit(-1);
49. }
50. firstPKRealStress.addMatrixProduct(0.0, *temp, invFt, detF);
51. }
52. else if (dimension == 3) {
53. (*temp)(0, 0) = (*thePDNodes[i]->cauchyStress)(0);
54. (*temp)(1, 1) = (*thePDNodes[i]->cauchyStress)(1);
55. (*temp)(2, 2) = (*thePDNodes[i]->cauchyStress)(2);
56. (*temp)(0, 1) = (*thePDNodes[i]->cauchyStress)(3);
57. (*temp)(0, 2) = (*thePDNodes[i]->cauchyStress)(4);
58. (*temp)(1, 2) = (*thePDNodes[i]->cauchyStress)(5);
59. (*temp)(1, 0) = (*temp)(0, 1);
60. (*temp)(2, 0) = (*temp)(0, 2);
61. (*temp)(2, 1) = (*temp)(1, 2);
62. Ft.addMatrixTranspose(0.0, *(thePDNodes[i]->defomation Gradi
    entF), 1.0);
63. Ft.Invert(invFt);
64. detF = (*(thePDNodes[i]->defomationGradientF))(0, 0)
    *(*(thePDNodes[i]-
    >defomationGradientF))(1,1)*(*(thePDNodes[i]-
    >defomationGradientF))(2,2) + (*(thePDNodes[i]-
    >defomationGradientF))(0, 1)*(*(thePDNodes[i]-
    >defomationGradientF))(1,2)*(*(thePDNodes[i]-
    >defomationGradientF))(2,0)+ (*(thePDNodes[i]-
    >defomationGradientF))(0, 2)*(*(thePDNodes[i]-
    >defomationGradientF))(1,0)*(*(thePDNodes[i]-
    >defomationGradientF))(2,1) - ((*(thePDNodes[i]-
```

```
>defomationGradientF))(0, 0)*(*(thePDNodes[i]->defomationGradientF))(1,2)*(*(thePDNodes[i]->defomationGradientF))(2, 1))- ((*(thePDNodes[i]->defomationGradientF))(0, 1)*(*(thePDNodes[i]->defomationGradientF))(1,0)*(*(thePDNodes[i]->defomationGradientF))(2, 2)) - ((*(thePDNodes[i]->defomationGradientF))(0, 2)*(*(thePDNodes[i]->defomationGradientF))(1,1)*(*(thePDNodes[i]->defomationGradientF))(2, 0));
65. if (abs(detF) <1e-14) {
66. opserr ≪" detF is zero, please check!" ≪endln;
67. }
68. firstPKRealStress.addMatrixProduct(0.0, *temp, invFt, detF);
69. }
70. for (int j = 0; j <= thePDNodes[i]->neighborNum; j++) {
71. const Vector &trialDispI = thePDNodes[i]->getTrialDisp();
72. const Vector &trialDispJ = thePDNodes[(*(thePDNodes[i]->theNeiborPDNodes))(j)]->getTrialDisp();
73. tempTrialDispI(0) = trialDispI(0) + (*perturb[i])(0);
74. tempTrialDispI(1) = trialDispI(1) + (*perturb[i])(1);
75. tempTrialDispJ(0) = trialDispJ(0) + (*perturb[(*(thePDNodes[i]->theNeiborPDNodes))(j)])(0);
76. tempTrialDispJ(1) = trialDispJ(1) + (*perturb[(*(thePDNodes[i]->theNeiborPDNodes))(j)])(1);
77. if (dimension == 3) {
78. tempTrialDispI(2) = trialDispI(2) + (*perturb[i])(2);
79. tempTrialDispJ(2) = trialDispJ(2) + (*perturb[(*(thePDNodes[i]->theNeiborPDNodes))(j)])(2);
80. }
81. tempForceState2.addMatrixVector(0.0, *(thePDNodes[i]->invShapeK), *(thePDNodes[i]->xi)[j], (*(thePDNodes[i]->weightFun))(j)*thePDNodes[(*(thePDNodes[i]->theNeiborPDNodes))(j)]->getVolume()*thePDNodes[i]->getVolume());
82. tempForceState.addMatrixVector(0.0, firstPKRealStress, tempForceState2, 1.0);
```

```
83. (*(thePDNodes[i]->forceState))(j, 0) = tempForceState(0);
84. (*(thePDNodes[i]->forceState))(j, 1) = tempForceState(1);
85. if (dimension == 3) {
86. (*(thePDNodes[i]->forceState))(j, 2) = tempForceState(2);
87. }
88. }
89. }
90. }
```

Among them, lines 13–28 calculate the Cauchy stress of the PD point using a material constitutive model (see Eq. (18.7)); lines 38–51 and lines 52–69 calculate the first PK stress of the PD point in the two-dimensional and three-dimensional, respectively (refer to Eq. (18.8)); lines 70–88 calculate the force state of the PD point (refer to Eq. (18.9)). Take the following example to describe the format of variable force-State: as shown in Fig. 18.7, the neighbor particles of #5 are [6, 7, 9, 8, 12], and (*(thePDNodes[4]->forceState)) stores all the force state of PD point #5. It is noted that #9 is the third neighbor particle of #5, and the force state $\underline{\mathbf{T}}[5]\langle\boldsymbol{\xi}_{59}\rangle$ in x direction can be written as (*(thePDNodes[4]->forceState))(3, 0).

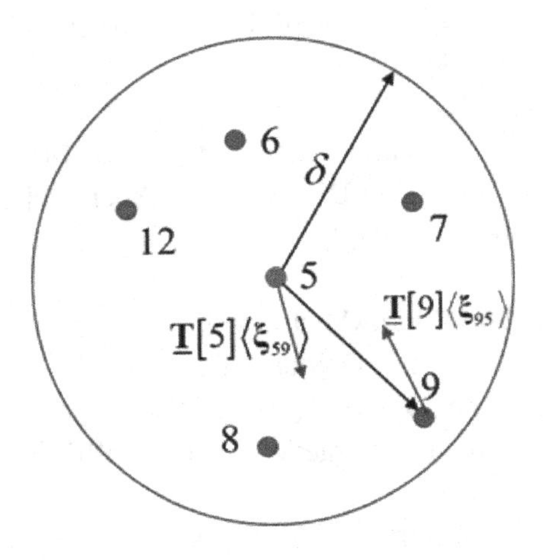

Fig. 18.7 The force state of PD point #5.

18.3 Application Examples

In this section, a 2D linear elastic column model under uniaxial compression is performed using static analysis. The geometry of the column is shown in Fig. 18.8. The width and length of the column are 1.02 m and 3.24 m, respectively. The column is discretized into 243 PD points, and the grid size dx is 0.12 m. A linear elastic material model with a plane strain condition is used with Young's modulus $E = 2.5e10$ Pa, and Poisson's ratio $v = 0.25$. The thickness t of the model is 0.12 m. The volume

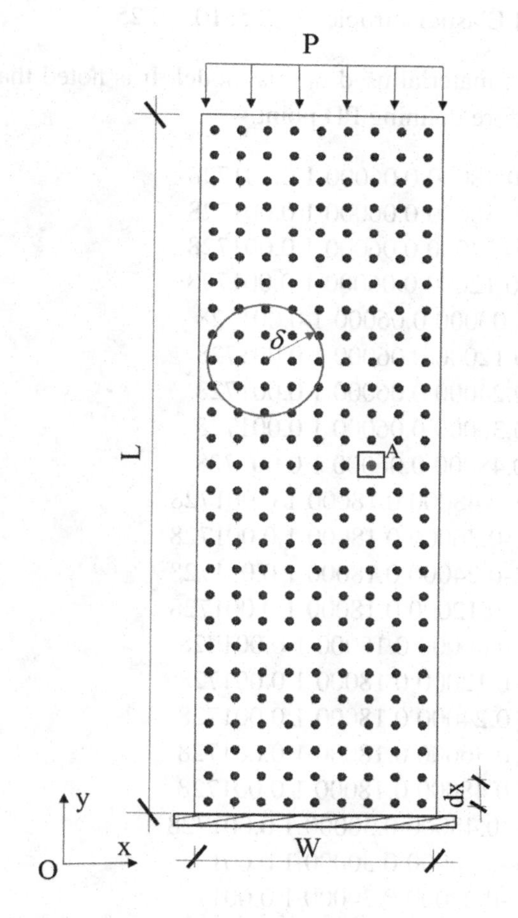

Fig. 18.8 The geometry of the column model.

for each PD point is $V = dx * dx * t$ (i.e., 0.001728 m^3). The uniaxial compressive load is assumed to be $P = 9.375e8$ Pa applied to the top boundary layer. At the bottom of the model, the degrees of freedom (DOFs) of the points are fixed in y direction and free in x direction except for the middle point. The quasi-static analysis is performed with time step $dt = 0.1$ s and total loading step $n = 10$.

The Tcl model is as follows:

```
1. wipe
2. model basic -ndm 2 -ndf 2
3. nDMaterial ElasticIsotropic 1   2.5e10   0.25
```

Line 3 set the material used in this model. It is noted that the material should be set before defining PD point.

```
4. pdnode 1 -0.48000 0.06000 1 0.001728
5. pdnode 2 -0.36000 0.06000 1 0.001728
6. pdnode 3 -0.24000 0.06000 1 0.001728
7. pdnode 4 -0.12000 0.06000 1 0.001728
8. pdnode 5 0.00000 0.06000 1 0.001728
9. pdnode 6 0.12000 0.06000 1 0.001728
10. pdnode 7 0.24000 0.06000 1 0.001728
11. pdnode 8 0.36000 0.06000 1 0.001728
12. pdnode 9 0.48000 0.06000 1 0.001728
13. pdnode 10 -0.48000 0.18000 1 0.001728
14. pdnode 11 -0.36000 0.18000 1 0.001728
15. pdnode 12 -0.24000 0.18000 1 0.001728
16. pdnode 13 -0.12000 0.18000 1 0.001728
17. pdnode 14 0.00000 0.18000 1 0.001728
18. pdnode 15 0.12000 0.18000 1 0.001728
19. pdnode 16 0.24000 0.18000 1 0.001728
20. pdnode 17 0.36000 0.18000 1 0.001728
21. pdnode 18 0.48000 0.18000 1 0.001728
22. pdnode 19 -0.48000 0.30000 1 0.001728
23. pdnode 20 -0.36000 0.30000 1 0.001728
24. pdnode 21 -0.24000 0.30000 1 0.001728
25. pdnode 22 -0.12000 0.30000 1 0.001728
```

```
26. pdnode 23 0.00000 0.30000 1 0.001728
27. pdnode 24 0.12000 0.30000 1 0.001728
28. pdnode 25 0.24000 0.30000 1 0.001728
29. pdnode 26 0.36000 0.30000 1 0.001728
30. pdnode 27 0.48000 0.30000 1 0.001728
31. pdnode 28 -0.48000 0.42000 1 0.001728
32. pdnode 29 -0.36000 0.42000 1 0.001728
33. pdnode 30 -0.24000 0.42000 1 0.001728
34. pdnode 31 -0.12000 0.42000 1 0.001728
35. pdnode 32 0.00000 0.42000 1 0.001728
36. pdnode 33 0.12000 0.42000 1 0.001728
37. pdnode 34 0.24000 0.42000 1 0.001728
38. pdnode 35 0.36000 0.42000 1 0.001728
39. pdnode 36 0.48000 0.42000 1 0.001728
40. pdnode 37 -0.48000 0.54000 1 0.001728
41. pdnode 38 -0.36000 0.54000 1 0.001728
42. pdnode 39 -0.24000 0.54000 1 0.001728
43. pdnode 40 -0.12000 0.54000 1 0.001728
44. pdnode 41 0.00000 0.54000 1 0.001728
45. pdnode 42 0.12000 0.54000 1 0.001728
46. pdnode 43 0.24000 0.54000 1 0.001728
47. pdnode 44 0.36000 0.54000 1 0.001728
48. pdnode 45 0.48000 0.54000 1 0.001728
49. pdnode 46 -0.48000 0.66000 1 0.001728
50. pdnode 47 -0.36000 0.66000 1 0.001728
51. pdnode 48 -0.24000 0.66000 1 0.001728
52. pdnode 49 -0.12000 0.66000 1 0.001728
53. pdnode 50 0.00000 0.66000 1 0.001728
54. pdnode 51 0.12000 0.66000 1 0.001728
55. pdnode 52 0.24000 0.66000 1 0.001728
56. pdnode 53 0.36000 0.66000 1 0.001728
57. pdnode 54 0.48000 0.66000 1 0.001728
58. pdnode 55 -0.48000 0.78000 1 0.001728
59. pdnode 56 -0.36000 0.78000 1 0.001728
60. pdnode 57 -0.24000 0.78000 1 0.001728
61. pdnode 58 -0.12000 0.78000 1 0.001728
```

```
62. pdnode 59 0.00000 0.78000 1 0.001728
63. pdnode 60 0.12000 0.78000 1 0.001728
64. pdnode 61 0.24000 0.78000 1 0.001728
65. pdnode 62 0.36000 0.78000 1 0.001728
66. pdnode 63 0.48000 0.78000 1 0.001728
67. pdnode 64 -0.48000 0.90000 1 0.001728
68. pdnode 65 -0.36000 0.90000 1 0.001728
69. pdnode 66 -0.24000 0.90000 1 0.001728
70. pdnode 67 -0.12000 0.90000 1 0.001728
71. pdnode 68 0.00000 0.90000 1 0.001728
72. pdnode 69 0.12000 0.90000 1 0.001728
73. pdnode 70 0.24000 0.90000 1 0.001728
74. pdnode 71 0.36000 0.90000 1 0.001728
75. pdnode 72 0.48000 0.90000 1 0.001728
76. pdnode 73 -0.48000 1.02000 1 0.001728
77. pdnode 74 -0.36000 1.02000 1 0.001728
78. pdnode 75 -0.24000 1.02000 1 0.001728
79. pdnode 76 -0.12000 1.02000 1 0.001728
80. pdnode 77 0.00000 1.02000 1 0.001728
81. pdnode 78 0.12000 1.02000 1 0.001728
82. pdnode 79 0.24000 1.02000 1 0.001728
83. pdnode 80 0.36000 1.02000 1 0.001728
84. pdnode 81 0.48000 1.02000 1 0.001728
85. pdnode 82 -0.48000 1.14000 1 0.001728
86. pdnode 83 -0.36000 1.14000 1 0.001728
87. pdnode 84 -0.24000 1.14000 1 0.001728
88. pdnode 85 -0.12000 1.14000 1 0.001728
89. pdnode 86 0.00000 1.14000 1 0.001728
90. pdnode 87 0.12000 1.14000 1 0.001728
91. pdnode 88 0.24000 1.14000 1 0.001728
92. pdnode 89 0.36000 1.14000 1 0.001728
93. pdnode 90 0.48000 1.14000 1 0.001728
94. pdnode 91 -0.48000 1.26000 1 0.001728
95. pdnode 92 -0.36000 1.26000 1 0.001728
96. pdnode 93 -0.24000 1.26000 1 0.001728
97. pdnode 94 -0.12000 1.26000 1 0.001728
```

98. pdnode 95 0.00000 1.26000 1 0.001728
99. pdnode 96 0.12000 1.26000 1 0.001728
100. pdnode 97 0.24000 1.26000 1 0.001728
101. pdnode 98 0.36000 1.26000 1 0.001728
102. pdnode 99 0.48000 1.26000 1 0.001728
103. pdnode 100 -0.48000 1.38000 1 0.001728
104. pdnode 101 -0.36000 1.38000 1 0.001728
105. pdnode 102 -0.24000 1.38000 1 0.001728
106. pdnode 103 -0.12000 1.38000 1 0.001728
107. pdnode 104 0.00000 1.38000 1 0.001728
108. pdnode 105 0.12000 1.38000 1 0.001728
109. pdnode 106 0.24000 1.38000 1 0.001728
110. pdnode 107 0.36000 1.38000 1 0.001728
111. pdnode 108 0.48000 1.38000 1 0.001728
112. pdnode 109 -0.48000 1.50000 1 0.001728
113. pdnode 110 -0.36000 1.50000 1 0.001728
114. pdnode 111 -0.24000 1.50000 1 0.001728
115. pdnode 112 -0.12000 1.50000 1 0.001728
116. pdnode 113 0.00000 1.50000 1 0.001728
117. pdnode 114 0.12000 1.50000 1 0.001728
118. pdnode 115 0.24000 1.50000 1 0.001728
119. pdnode 116 0.36000 1.50000 1 0.001728
120. pdnode 117 0.48000 1.50000 1 0.001728
121. pdnode 118 -0.48000 1.62000 1 0.001728
122. pdnode 119 -0.36000 1.62000 1 0.001728
123. pdnode 120 -0.24000 1.62000 1 0.001728
124. pdnode 121 -0.12000 1.62000 1 0.001728
125. pdnode 122 0.00000 1.62000 1 0.001728
126. pdnode 123 0.12000 1.62000 1 0.001728
127. pdnode 124 0.24000 1.62000 1 0.001728
128. pdnode 125 0.36000 1.62000 1 0.001728
129. pdnode 126 0.48000 1.62000 1 0.001728
130. pdnode 127 -0.48000 1.74000 1 0.001728
131. pdnode 128 -0.36000 1.74000 1 0.001728
132. pdnode 129 -0.24000 1.74000 1 0.001728
133. pdnode 130 -0.12000 1.74000 1 0.001728

```
134. pdnode 131 0.00000 1.74000 1 0.001728
135. pdnode 132 0.12000 1.74000 1 0.001728
136. pdnode 133 0.24000 1.74000 1 0.001728
137. pdnode 134 0.36000 1.74000 1 0.001728
138. pdnode 135 0.48000 1.74000 1 0.001728
139. pdnode 136 -0.48000 1.86000 1 0.001728
140. pdnode 137 -0.36000 1.86000 1 0.001728
141. pdnode 138 -0.24000 1.86000 1 0.001728
142. pdnode 139 -0.12000 1.86000 1 0.001728
143. pdnode 140 0.00000 1.86000 1 0.001728
144. pdnode 141 0.12000 1.86000 1 0.001728
145. pdnode 142 0.24000 1.86000 1 0.001728
146. pdnode 143 0.36000 1.86000 1 0.001728
147. pdnode 144 0.48000 1.86000 1 0.001728
148. pdnode 145 -0.48000 1.98000 1 0.001728
149. pdnode 146 -0.36000 1.98000 1 0.001728
150. pdnode 147 -0.24000 1.98000 1 0.001728
151. pdnode 148 -0.12000 1.98000 1 0.001728
152. pdnode 149 0.00000 1.98000 1 0.001728
153. pdnode 150 0.12000 1.98000 1 0.001728
154. pdnode 151 0.24000 1.98000 1 0.001728
155. pdnode 152 0.36000 1.98000 1 0.001728
156. pdnode 153 0.48000 1.98000 1 0.001728
157. pdnode 154 -0.48000 2.10000 1 0.001728
158. pdnode 155 -0.36000 2.10000 1 0.001728
159. pdnode 156 -0.24000 2.10000 1 0.001728
160. pdnode 157 -0.12000 2.10000 1 0.001728
161. pdnode 158 0.00000 2.10000 1 0.001728
162. pdnode 159 0.12000 2.10000 1 0.001728
163. pdnode 160 0.24000 2.10000 1 0.001728
164. pdnode 161 0.36000 2.10000 1 0.001728
165. pdnode 162 0.48000 2.10000 1 0.001728
166. pdnode 163 -0.48000 2.22000 1 0.001728
167. pdnode 164 -0.36000 2.22000 1 0.001728
168. pdnode 165 -0.24000 2.22000 1 0.001728
169. pdnode 166 -0.12000 2.22000 1 0.001728
```

```
170. pdnode 167 0.00000 2.22000 1 0.001728
171. pdnode 168 0.12000 2.22000 1 0.001728
172. pdnode 169 0.24000 2.22000 1 0.001728
173. pdnode 170 0.36000 2.22000 1 0.001728
174. pdnode 171 0.48000 2.22000 1 0.001728
175. pdnode 172 -0.48000 2.34000 1 0.001728
176. pdnode 173 -0.36000 2.34000 1 0.001728
177. pdnode 174 -0.24000 2.34000 1 0.001728
178. pdnode 175 -0.12000 2.34000 1 0.001728
179. pdnode 176 0.00000 2.34000 1 0.001728
180. pdnode 177 0.12000 2.34000 1 0.001728
181. pdnode 178 0.24000 2.34000 1 0.001728
182. pdnode 179 0.36000 2.34000 1 0.001728
183. pdnode 180 0.48000 2.34000 1 0.001728
184. pdnode 181 -0.48000 2.46000 1 0.001728
185. pdnode 182 -0.36000 2.46000 1 0.001728
186. pdnode 183 -0.24000 2.46000 1 0.001728
187. pdnode 184 -0.12000 2.46000 1 0.001728
188. pdnode 185 0.00000 2.46000 1 0.001728
189. pdnode 186 0.12000 2.46000 1 0.001728
190. pdnode 187 0.24000 2.46000 1 0.001728
191. pdnode 188 0.36000 2.46000 1 0.001728
192. pdnode 189 0.48000 2.46000 1 0.001728
193. pdnode 190 -0.48000 2.58000 1 0.001728
194. pdnode 191 -0.36000 2.58000 1 0.001728
195. pdnode 192 -0.24000 2.58000 1 0.001728
196. pdnode 193 -0.12000 2.58000 1 0.001728
197. pdnode 194 0.00000 2.58000 1 0.001728
198. pdnode 195 0.12000 2.58000 1 0.001728
199. pdnode 196 0.24000 2.58000 1 0.001728
200. pdnode 197 0.36000 2.58000 1 0.001728
201. pdnode 198 0.48000 2.58000 1 0.001728
202. pdnode 199 -0.48000 2.70000 1 0.001728
203. pdnode 200 -0.36000 2.70000 1 0.001728
204. pdnode 201 -0.24000 2.70000 1 0.001728
205. pdnode 202 -0.12000 2.70000 1 0.001728
```

```
206. pdnode 203 0.00000 2.70000 1 0.001728
207. pdnode 204 0.12000 2.70000 1 0.001728
208. pdnode 205 0.24000 2.70000 1 0.001728
209. pdnode 206 0.36000 2.70000 1 0.001728
210. pdnode 207 0.48000 2.70000 1 0.001728
211. pdnode 208 -0.48000 2.82000 1 0.001728
212. pdnode 209 -0.36000 2.82000 1 0.001728
213. pdnode 210 -0.24000 2.82000 1 0.001728
214. pdnode 211 -0.12000 2.82000 1 0.001728
215. pdnode 212 0.00000 2.82000 1 0.001728
216. pdnode 213 0.12000 2.82000 1 0.001728
217. pdnode 214 0.24000 2.82000 1 0.001728
218. pdnode 215 0.36000 2.82000 1 0.001728
219. pdnode 216 0.48000 2.82000 1 0.001728
220. pdnode 217 -0.48000 2.94000 1 0.001728
221. pdnode 218 -0.36000 2.94000 1 0.001728
222. pdnode 219 -0.24000 2.94000 1 0.001728
223. pdnode 220 -0.12000 2.94000 1 0.001728
224. pdnode 221 0.00000 2.94000 1 0.001728
225. pdnode 222 0.12000 2.94000 1 0.001728
226. pdnode 223 0.24000 2.94000 1 0.001728
227. pdnode 224 0.36000 2.94000 1 0.001728
228. pdnode 225 0.48000 2.94000 1 0.001728
229. pdnode 226 -0.48000 3.06000 1 0.001728
230. pdnode 227 -0.36000 3.06000 1 0.001728
231. pdnode 228 -0.24000 3.06000 1 0.001728
232. pdnode 229 -0.12000 3.06000 1 0.001728
233. pdnode 230 0.00000 3.06000 1 0.001728
234. pdnode 231 0.12000 3.06000 1 0.001728
235. pdnode 232 0.24000 3.06000 1 0.001728
236. pdnode 233 0.36000 3.06000 1 0.001728
237. pdnode 234 0.48000 3.06000 1 0.001728
238. pdnode 235 -0.48000 3.18000 1 0.001728
239. pdnode 236 -0.36000 3.18000 1 0.001728
240. pdnode 237 -0.24000 3.18000 1 0.001728
241. pdnode 238 -0.12000 3.18000 1 0.001728
242. pdnode 239 0.00000 3.18000 1 0.001728
```

```
243. pdnode 240 0.12000 3.18000 1 0.001728
244. pdnode 241 0.24000 3.18000 1 0.001728
245. pdnode 242 0.36000 3.18000 1 0.001728
246. pdnode 243 0.48000 3.18000 1 0.001728
```

Lines 4–246 define the PD points, including the tag, coordinates, material and volume of each PD point.

```
247. set startNum 1
248. set lastNum 243
249. set dx 0.120000
250. fix 1 0 1
251. fix 2 0 1
252. fix 3 0 1
253. fix 4 0 1
254. fix 5 1 1
255. fix 6 0 1
256. fix 7 0 1
257. fix 8 0 1
258. fix 9 0 1
```

Lines 247–258 represent constraints. The DOFs of PD points from 1 to 9 in *y* direction are fixed and free in x direction except for the middle point.

```
259. element stateBasedPeridynamics 1 2 $startNum $lastNum [expr
     2.015*$dx] $dx -thickness 0.12
```

Line 259 defines the element "stateBasedPeridynamics" using the dimension, the tags of the first and the last points, the horizon size, the grid size and the thickness.

```
260. recorder Node -file displacementPD.out -time -nodeRange $startNum
     $lastNum -precision 16 -dof 1 2 disp
261. recorder Node -file nodeReactionPD.out -time -nodeRange 1 10 -dof
     1 2 reaction
262. recorder Element -file stressPDNode.out -time -ele 1 stress
263. recorder Element -file strainPDNode.out -time -ele 1 strain
```

Lines 260–263 are recorders. The displacement, stress and strain of each PD point are recorded.

```
264. pattern Plain 1 Linear {
265. load 235 0.000000000000000 -13500000
266. load 236 0.000000000000000 -13500000
267. load 237 0.000000000000000 -13500000
268. load 238 0.000000000000000 -13500000
269. load 239 0.000000000000000 -13500000
270. load 240 0.000000000000000 -13500000
271. load 241 0.000000000000000 -13500000
272. load 242 0.000000000000000 -13500000
273. load 243 0.000000000000000 -13500000
274. }
275. constraints Transformation
276. numberer RCM
277. system BandGeneral
278. test NormDispIncr 1.0e-6 50 1
279. algorithm Newton
280. integrator LoadControl 0.1
281. analysis Static
282. analyze 10
```

Lines 264–282 perform the static analysis for the column under uniaxial compression loading condition.

The computed results are compared with those calculated using the FEM, which discretizes the model into 208 quad elements defined as displacement-based four-node finite elements with four integration points each. The material model and boundary conditions are all similar to the PD model.

Figure 18.9 shows the comparison of the displacement calculated using SPD and FEM. The results show that SPD is in good agreement with those of FEM. Figure 18.10 shows horizontal and vertical displacement histories of representative points A; it shows that the SPD is not accurate since the error gradually increases as the applied force increases. This

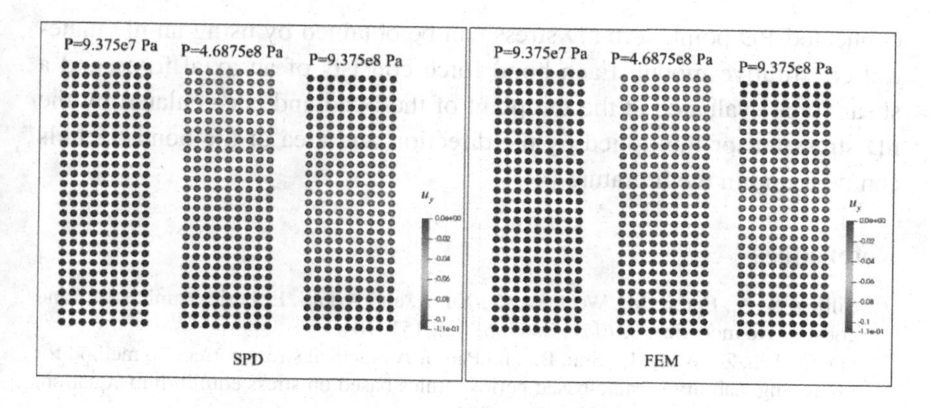

Fig. 18.9 The displacements predicted PD and FEM.

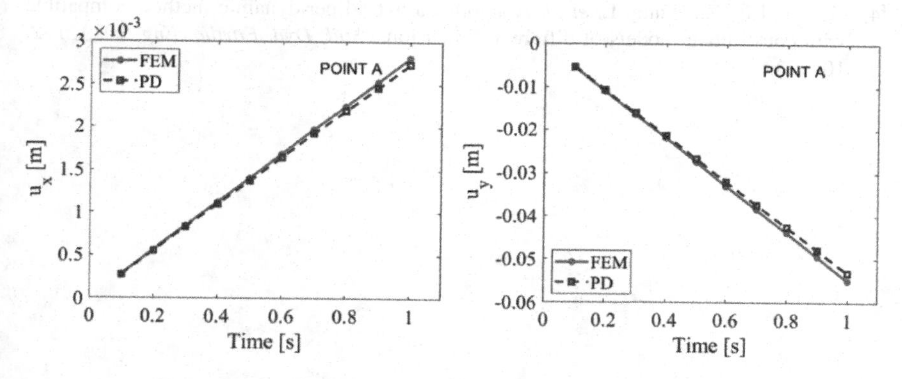

Fig. 18.10 Horizontal and vertical displacement histories of representative points A using PD and FEM.

is mainly caused by the so-called zero-energy mode in SPD. There are several methods to solve the inaccurate and unstable problems, e.g., the authors have proposed a stress correction method to mitigate the zero-energy mode [2, 3], which is implemented based on the element introduced in this section. Recently, a stabilized hybrid PD (HPD) method has been presented whose framework is similar to BPD. However, the HPD calculates a 2D or 3D (or nD) strain at the middle point of each bond by PD points in a new subhorizon defined by the overlap of two horizons of the connected PD points. An nD stress can be obtained by using an nD material constitutive model. Each bond force consists of an axial force and a shear force evaluated at the midpoint of the bond and is calculated by the nD stress tensor multiplied by the direction and area of the bond. Details can be found in the literature [4].

References

[1] Silling, S. A., Epton, M., Weckner, O., Xu, J., and Askari, E. Peridynamic states and constitutive modeling. *J. Elast.* 2007; *88* (2): 151–184.

[2] Gu, Q., Lin, Z., Wang, L., Sun, B., and Pan, J. A practical stress correction method for improving stability of state-based peridynamics based on stress equilibrium equation. *Int. J. Struct. Stab. Dyn.* 2022; *22* (8): 2250094.

[3] Wang, L., Huang, S., Gu, Q. *et al.* Simulation of highly nonlinear materials based on a stabilized non-ordinary state-based peridynamic model. *Soil. Dyn. Earthq. Eng.* 2022, *157*: 107250.

[4] Gu, Q., Lin, Z., Wang, L. *et al.* A stabilized hybrid peridynamic method compatible with constitutive models of different dimensions. *Soil. Dyn. Earthq. Eng.* 2023, *172*: 107903.

Index

Printed in the United States
by Baker & Taylor Publisher Services

Printed in the United States
by Baker & Taylor Publisher Services